APPLICATION OF BRAID GROUPS
IN 2D HALL SYSTEM PHYSICS

COMPOSITE FERMION STRUCTURE

APPLICATION OF BRAID GROUPS IN 2D HALL SYSTEM PHYSICS
COMPOSITE FERMION STRUCTURE

Janusz Jacak • Ryszard Gonczarek
Lucjan Jacak • Ireneusz Jóźwiak

Wroclaw University of Technology, Poland

 **World Scientific**

NEW JERSEY • LONDON • SINGAPORE • BEIJING • SHANGHAI • HONG KONG • TAIPEI • CHENNAI

Published by

World Scientific Publishing Co. Pte. Ltd.

5 Toh Tuck Link, Singapore 596224

USA office: 27 Warren Street, Suite 401-402, Hackensack, NJ 07601

UK office: 57 Shelton Street, Covent Garden, London WC2H 9HE

British Library Cataloguing-in-Publication Data
A catalogue record for this book is available from the British Library.

APPLICATION OF BRAID GROUPS IN 2D HALL SYSTEM PHYSICS
Composite Fermion Structure

ISBN-13 978-981-4412-02-5
ISBN-10 981-4412-02-3

Printed in Singapore by World Scientific Printers.

Preface

The topological methods for the description of quantum many particle systems are presented within the braid group approach. 2D multi-particle systems are of special interest because of the exceptional abundance of braid group structure for planar systems, which produce a rich 2D Hall system physics in the presence of strong magnetic fields. A new topological method is proposed in order to explain the Laughlin correlations in 2D charged systems in a magnetic field, corresponding to the fractional filling of the lowest Landau level. The reason of fractional quantum Hall effect peculiarity is associated within the new approach with the situation when cyclotron trajectories are too short in comparison to the particle separation, which precludes particle exchanges along cyclotron trajectories, although these exchanges are required for the definition of the statistics, according to the braid group scheme of quantization. Such a situation happens at fractional fillings of the lowest Landau level. The cyclotron braid subgroups of the full braid group, comprised of multi-loop trajectories are originally identified in order to restore the exchanges of particles and to allow statistics determination when single-loop cyclotron orbits cannot reach neighboring particles. The statistics of quantum particles is associated with a one-dimensional unitary representation selected for the defined cyclotron subgroups upon the path integral formulation for not simply connected configuration spaces. In this manner, the structure of composite fermions linked to Laughlin correlations can be elucidated without invoking any auxiliary fictitious elements, such as flux-tubes or vortices, used in the traditional heuristic modeling of composite fermions. This model has been connected to the standard theory of composite fermions including a derivation of the fractional quantum Hall effect hierarchy, via mapping of the fractional Hall effect onto the integer Hall effect within the formulated cyclotron braid theory. Arguments supporting the cyclotron braid approach are collected, with a special attention paid to the recent experimental data for suspended graphene, which indicate that carrier mobility plays a crucial role in the competition between Laughlin collective state and insulating Wigner-crystal-type state. Some linkages to the newest research in the field of topological insulators and of 2D topological states modeled in optical lattices are outlined. The wider presentation of braid group methods and their applications is also reviewed, including the braid group approach to both indistinguishable (quantum) and distinguishable (classical) identical particle systems on various manifolds.

Acknowledgments

Related research project supported by the Polish National Science Center (NCN), decision No. DEC-2011/02/A/ST3/001.

Contents

Chapter 1

Introduction

Global topological effects are increasingly seen as fundamental to physics. Advanced methods of largely topological character are widely applied in the domain of relativistic physics, field theory, and lately in attempts at formulating a description of quantum gravitation. Essentially, the basic aspects of electrodynamics with sources–charges and vortices–currents have geometrical conditioning, these both field defects have a locally indelible topological character. The study of topological defects (e.g., generalizations of the defect of a tape twisted at the ends, the so-called Sine–Gordon kink) was developed within field theory [Ryder (1996); Polyakov (1974); t'Hooft (1974)], leading to the concept of objects with a topological charge (e.g., instantons [t'Hooft (1976)]). Condensed matter physics also widely involves topological methods that are used e.g., to characterize defects in crystals [Mermin (1979)], as well as defects and textures in quantum liquids with rich multidimensional order parameters (as in the cases of superfluid He^3 and liquid crystals) [Mermin (1979)]. The known global topological-type effects such as the Aharonov–Bohm effect [Aharonov and Bohm (1959)], the entire class of effects linked to Josephson superconducting circuits [Josephson (1974)], or even quantum interference phenomena such as Anderson localization [Anderson (1958)] apart from their fundamental character, have allowed the construction of some setups with practical significance, e.g., superconducting interferometer called SQUID (Superconducting Quantum Interference Device), which allows single quanta of magnetic flux to be registered.

Topological methods have proven to be extremely important in a very basic aspect, i.e., in the quantization of systems containing many identical particles especially in a 2D Hall geometry. The experimental discovery of the fractional quantum Hall effect (FQHE) [Tsui et al. (1982)] contributed

1

to the recognition of the role of spatial topology (of the manifold in which the system is located) in determining the characteristics of quantum particles undergoing dynamics in these spaces. The extreme topological richness of the physics of many particle systems in two-dimensional spaces was observed and expressed in the ability to carry out statistics different than those of fermions and bosons (namely, fractional anyon statistics).

The role of topological degrees of freedom in quantum topological information theory [Nielsen and Chuang (2000); Nayak *et al.* (2008); Preskill (2004)] has recently been considered in the effort to overcome or evade the naturally local decoherence[1], by locating the processing of information at global topological degrees of freedom, which are resistant to local decoherential deformations. The related scheme of so-called non-abelian anyons [Kitaev (2003); Nayak *et al.* (2008); Preskill (2004)] is promising, but it remains difficult in terms of practical implementation (controlling the topological degrees of freedom e.g., in Hall systems or superconductive systems seems far out of reach for present technology).

Algebraic topology plays the central role in the topological methods that are useful in physics [Spanier (1966)]. Algebraic topology is a broad field that describes the deformative and continuous character of mapping in various spaces, and it is strongly related to physical concepts such as e.g., configuration space trajectories for various physical spaces (that is, manifolds reflecting dynamics restrictions or boundary conditions), surfaces and their singularities (e.g., Fermi isoenergetic surfaces), spatial distributions of various fields and order parameters, or defects in their distribution.

The primary tools in algebraic topology are homotopy groups, which are called braid groups in the case of the configuration spaces for systems of many particles. Braid groups are basic [fundamental] groups (i.e., first homotopy groups π_1) of configuration spaces of multiparticle systems [Artin (1947); Birman (1974)]. Braid groups have been studied in-depth since the 1970s, and they have also been considered in the context of the related knot theory [Birman (1974)]. In current mathematics, significant attention has been directed toward algebraic topology and its relationships with other areas in mathematics.

Braid groups are the first homotopy groups i.e., the π_1 groups of multiparticle configuration spaces. The π_1 homotopy group is the set (with a group structure) of all classes of topologically nonequivalent trajectories

---

[1]Decoherence leading to uncontrolled leakage of quantum information is the major obstacle on the way to the construction of a large serviceable quantum computer [Bouwmeester *et al.* (2000); Jacak *et al.* (2009b, 2005)].

(i.e., trajectories that cannot be continuously transformed into one another by means of deformation, so that they are nonhomotopic) [2]. Braid groups are strongly dependent on the type of physical space in which the particle trajectories are located; these spaces are called manifolds. For example, R^3 is a three-dimensional space, R^2 is a surface, S^2 is a three-dimensional sphere, and T is a torus (in particular, the latter one corresponds to a rectangular plaquette with Born-Karman periodic boundary conditions imposed). Surprisingly, the braid groups for a surface (and also for a locally 2D manifold, such as a sphere or a torus) are complicated infinite groups, while they are finite and equal to a simple permutation group for a three-dimensional space (and for higher dimensional spaces as well). This significant difference can be attributed to the rich physics of two-dimensional systems, whose quantum properties are observed within the experimental physics of planar Hall systems and are inaccessible in higher dimensions.

Despite the fact that since the study of Hall systems began in the early 1980s, it was clear that the topological dissimilarity of planar systems is expressed in the form of exotic and exceptionally rich physics, especially in the presence of strong magnetic fields, identification of the l whole topological conditioning has not been achieved. The subject matter of this treatise is the attempt at supplement this topological approach. Through the original definition of cyclotron braid groups, we demonstrate the recovery of the Laughlin correlations in charged multiparticle 2D systems in a strong perpendicular magnetic field. In this way, it seems, that we have identified the cause of exotic two-dimensional Laughlin correlations, which have been previously addressed in phenomenological categories. We have explained the nature of the auxiliary flux tubes of the field flux quanta, which are attached to particles in the phenomenological models of composite fermions that were constructed to illustrate Laughlin correlations and in order to develop effective calculation methods. The topological explanation of the structure of vortices in other phenomenological models of composite fermions has been found in a similar way.

This treatise is organized as follows. Chapter two presents a synthesis of the main aspects of Hall system physics and Laughlin correlations. Chapter three presents an outline of topological methods and of the braid group theory. In chapter four, the cyclotron braid subgroups used for the topological identification of the conditioning for Laughlin correlations are

---

[2]In a similar way, we can consider the homotopy of surfaces and higher dimensional hypersurfaces, which leads to homotopy groups π_2, π_3, etc. [Birman (1974); Mermin (1979)].

defined. In several subsections, various aspects of the proposed approach and the developed formalism are presented. Chapter five is addressed to recent progress in the experiment in Hall systems and to novel concept of so-called topological insulators. The conclusions summarize the main thesis: the explanation of Laughlin correlations and the structure of composite fermions in terms of multi-loop cyclotron braids, which occur naturally and unavoidable in planar multiparticle systems at sufficiently strong magnetic fields. Chapter six contains selected addenda and broader discussions of the previously raised issues, as well as a description of the novel concept of using pure braid groups to code information.

Chapter 2

Elements of Hall system physics in 2D spaces

One of the major discoveries in physics in the late 20th century was the observation [Tsui *et al.* (1982)] of the fractional quantum Hall effect (FQHE, *Fractional Quantum Hall Effect*) in multiparticle systems of charged particles limited to two-dimensional dynamics in the presence of a strong perpendicular magnetic field, corresponding to fractional filling of the lowest Landau level (LLL, cf. Chapter 7.1). This unusual experiment provided the opportunity to observe the realization of different quantum mechanics in relation to the well-known quantum mechanics of fermions or bosons in 3D space. The peculiar behavior of the two-dimensional electron liquid did not fit into the framework of the integer quantum Hall effect (IQHE, *Integer Quantum Hall Effect*), which was previously observed experimentally [von Klitzing *et al.* (1980)] in a 2D system and was understandable in terms of the filling of subsequent Landau levels. The previously widely investigated Shubnikov–de Haas effect (in relation to the conductivity quantization) and de Haas–van Alphen effect (in reference to the quantization of magnetization [1]) [Abrikosov *et al.* (1975)], which were typically observed in experiments in 3D metals, actually belong to the same class of effects. Although it is analogous to the Shubnikov–de Haas effect, the observation of the IQHE in 2D systems refers to quasi-two-dimensional systems in which the third direction (along the field) did not introduce a continuous variable, which would slightly blur the quantization in the perpendicular plane to the magnetic field, as in 3D systems. The discovery of the FQHE definitely opened a new field of quantum physics that had no analogy in 3D systems. Shortly after the experimental discovery [Tsui *et al.* (1982)], the effect was

---

[1]The effects involve passing the Fermi surface through discrete Landau levels (as a result of their increasing degeneracy as the magnetic field increases in amplitude), which leads to oscillations of the magnetization or conductivity with field variation.

explained by R. B. Laughlin [Laughlin (1983b,a)], who proposed a suitable interpretation of the observed effect in terms of the famous quantum correlations in 2D, expressed in terms of the Laughlin wave function. Both the originators of the FQHE experiment (D. C. Tsui & H. L. Störmer) and the author of a proper explanation of the effect (R. B. Laughlin) were awarded the Nobel Prize in physics in 1998 (earlier, in 1985, the Nobel Prize was also awarded to K. von Klitzing for experimental observation of the IQHE in a 2D system).

The core aspect of the FQHE discovery was to demonstrate the different realization of quantum mechanics in 2D spaces, which has no equivalent in 3D. Although the Coulomb interaction of charged particles on a plane plays a major role in generating the effect, the role of this interaction cannot be explained in terms of particle dressing with interaction in the scheme of quasiparticle formation, which is well-known in 3D condensed matter physics. The specific role of the Coulomb interaction in the case of 2D is expressed by a 'distance quantization', as originally described by Laughlin [Laughlin (1983b)]. The matrix element of the interaction is not a continuous function of distance expressed in terms of the relative angular momentum of a pair of particles [Laughlin (1983a)]. This situation is illustrated by a presentation of the Coulomb interaction in the form of the so-called Haldane pseudopotentials [Girvin *et al.* (1985); Girvin and MacDonald (1987); Haldane (1983)], which express the projection of the interaction onto subspaces of the pairs' relative angular momentum. The Laughlin function of order p (for filling $\frac{1}{p}$ LLL, where $p = 3, 5...$) turns out to be the exact (not variational) state, provided the interaction is limited to only pseudopotentials with a relative angular momentum that is not greater than $p - 2$ [Haldane (1983)]. These pseudopotentials correspond to a short-range part of interaction, which appears to be crucial to the generation of Laughlin correlations.

2.1 Laughlin function

The Laughlin function [Laughlin (1983b)] is a simple generalization of the Slater function. The latter one is assumed for the full filling of the LLL (i.e., for $p = 1$) when none of the Haldane pseudopotentials plays a role (because $p - 2 < 0$), and the function of non-interacting fermions in a magnetic field is the exact solution, which has the form of the Vandermonde polynomial with the Gaussian exponential factor [Landau and Lifshitz (1972)]:

$$\Psi_S(z_1, ... z_N) = \prod_{i,j=1, i>j}^{N} (z_i - z_j) e^{-\sum_{i=1}^{N} \frac{|z_i|^2}{4l^2}}, \tag{2.1}$$

where $z_i = x_i + y_i$ is a complex representation of the position of the i-th particle on a plane and $l = \sqrt{\frac{\hbar c}{eB}}$ is a 'magnetic' distance scale (magnetic length).

The Laughlin function has the following form:

$$\Psi_L(z_1, ... z_N) = \prod_{i,j=1, i>j}^{N} (z_i - z_j)^p e^{-\sum_{i=1}^{N} \frac{|z_i|^2}{4l^2}}. \tag{2.2}$$

The Laughlin function, similar to the Slater function, is an anti-symmetric function (for an odd integer q); however, unlike the Slater function, when the particles are interchanging their positions on a plane, the Laughlin function gains a phase shift equal to $p\pi$, rather than the phase shift of π obtained for the Slater function, and although $e^{ip\pi} = e^{i\pi} = -1$, this $p\pi$ phase shift marks the Laughlin correlations. The factor of $\prod_{i,j=1, i>j}^{N}(z_i - z_j)^p$ in the Laughlin function is a Jastrow polynomial, and its multipliers, $(z_i - z_j)^p$, can be interpreted as p-fold zeros pinned to the particles when $z_i = z_j$ (described here as 'attaching p-fold zeros' to particles [Read (1994)]).

The Laughlin function is guessed, and there are no arguments that indicate why there are correlations expressed by the Jastrow factor for the filling of $\frac{1}{p}$ of the form [2] $\prod_{i,j=1, i>j}^{N}(z_i - z_j)^p$. However, there seems to be a topological reason that the Laughlin correlations are characteristic of 2D topology, but they are not retained in 3D (or in higher dimensions). The purpose of this treatise is to identify the cause of this topological distinction and recover the Laughlin correlations based on deeper topological conditions.

2.2 Composite fermions

2.2.1 *Composite fermions in Jain's model*

The phase shift $p\pi$ of the Laughlin function that is encountered while replacing particles on a plane has been accounted for within the phenomeno-

---

[2]It is shown, however, that the Laughlin function of degree p corresponds to the LLL filling of $\frac{1}{p}$; the proof is based on the similarity of the Laughlin function module to the equilibrium thermodynamic distribution of charged particles at the background with a charge-density equal to $\frac{1}{p}$ [Laughlin (1983a)], which expresses the role that the Coulomb interaction plays in generating the Laughlin states.

logical model of composite fermions [Jain (1989, 2007); Heinonen (1998)], which are phenomenologically considered as fermions with local flux tubes of the magnetic field flux quanta attached to each particle. If we follow Jain and assume that a local field flux tube equals to an even number q of flux quanta, $q\frac{hc}{e}$, then after replacing the particles with flux tubes attached to them we will we will get a phase shift of $(q+1)\pi$ due to the Aharonov–Bohm effect [Aharonov and Bohm (1959); Wilczek (1990)], which corresponds to the requirements of the Laughlin correlations. The concept of fictitious flux tubes is very useful; in particular, it allows restoring the main series of the so-called hierarchy of LLL fillings, at which the FQHE occurs [Jain (1989)]. This hierarchy can easily be identified if we consider that the local flux tubes of composite fermions in the mean field approximation may reduce the external magnetic field, leading to the IQHE in the resultant field. The IQHE states in the resultant field correspond to subsequent completely filled Landau levels and are interpreted as incompressible (because of the energy gap between the subsequent filled Landau levels in the resultant field [3]) FQHE states at LLL fillings, $\nu = \frac{n}{2n\pm1}$, where $\pm$ corresponds to the two possible orientations of the resultant field with respect to the external field and n corresponds to the number of filled Landau levels in the resultant field. This appealing model reduces the FQHE to a Shubnikov–de Haas oscillations [Abrikosov (1972)] in the residual field produced by screening of the true external field corresponding to the filling of $\nu = \frac{1}{2}$, at which the external field is completely reduced to zero by the mean field of the flux tubes (in case of $p = 3$)[4]. The Hall metal and the states within the FQHE hierarchy (near the Hall metal), which may be interpreted in terms of Shubnikov–de Haas oscillations, have been observed in increasingly precise experiments[5], which is clearly visible in the attached figure (Fig. 2.1) after the paper [Pan *et al.* (2003)].

---

[3]Reducing the sample surface increases the density of particles and leads to growth of the screening field, which is blocked by the energy gap.

[4]Therefore, for $\nu = \frac{1}{2}$, we get a resultant magnetic field with an effective magnitude of zero, despite the strong true external field. This effect produces a Fermi liquid state, which is called a Hall metal in this context. The situation is also repeated for other fillings with larger even denominators and related larger numbers of flux tube quanta attached to the particles.

[5]Even in relatively weaker magnetic fields (~ 10 T) compared to those of former experiments (~ 20 T), for continuously improving samples with high carrier density and mobility.

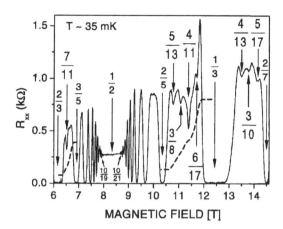

Fig. 2.1 Hall conductivity features corresponding to FQHE in a GaAs/AlGaAs quantum well with electron density of 10^{11} cm^{-2}. R_{xx} for $\frac{2}{3} > \nu > \frac{2}{7}$ at a temperature of $T \sim 35$ mK. Major fractions have been marked with arrows. The Hall resistance R_{xy} in the region of $\nu = \frac{7}{11}$ and $\nu = \frac{4}{11}$ is marked with a dotted line. *Source: W. Pan, H. L. Störmer, D. C. Tsui, L. N. Pfeiffer, K. W. Baldwin, and K. W. West, "Fractional quantum Hall effect of composite fermions," Phys. Rev. Lett. 90, p. 016801, 2003.*

2.2.2 Composite fermions in Read's model

A different approach to composite particles (in particular, to composite fermions) was formulated later, with the use of a collective object: a vortex attached to particles [Read (1994, 1989)]. Vortices were introduced there in analogy to vortices in superfluids, in terms of the Ginzburg–Landau functional, and they were used to describe the FQHE [Halperin *et al.* (1993)]. The elementary vortex with its center at a point z on a plane has been defined as follows [Read (1994)]:

$$V(z) = \prod_{j=1}^{N} (z_j - z),\qquad (2.3)$$

Whereas the vortex with a vorticity of q and a center z is defined by the formula:

$$V'(z) = \prod_{j=1}^{N} (z_j - z)^q.\qquad (2.4)$$

Attaching vortices to particles is interpreted as the replacement of z with z_i ($i \neq j$) and the addition of the i index to the product; as a result, we obtain

$$V(z) = \prod_{i,j=1, i>j}^{N} (z_j - z_i) \tag{2.5}$$

and

$$V'(z) = \prod_{i,j=1, i>j}^{N} (z_j - z_i)^q, \tag{2.6}$$

for elementary vortices and vortices with vorticity of q, respectively. The Vandermonde and Jastrow polynomials can be identified in the expressions (2.5) and (2.6), correspondingly. Therefore, the vortices can be expected to reflect peculiar Laughlin correlations because they are actually factors contributing to the Laughlin function. In addition to the phase element (expressed by the vorticity and equal to the phase shift caused by localized magnetic flux tubes in Jain's model), the vortices include the radial dependence of the Laughlin function in their definition (through the radial dependence of the Jastrow polynomial), which is related to the minimization of the Coulomb energy (expressed in terms of Haldane pseudopotentials). The concept of Read's vortices has proven to be useful because of their unusual properties, particularly the depletion of the local charge density and the resultant charge screening (by the positive charge of the *jellium*) of composite particles, as well as their capability to modify the vortex structure by shifting its center with respect to the electron related to it. The latter is used in the modeling of state functions other than the Laughlin function (e.g., for filling $\nu = \frac{5}{2}$ in the form of the so-called Pfaffian) [6] when the BCS pairing of fermions (expressed in position space by the Pfaffian, cf. Chapter 7.3) occurs as a result of Fermi sea instability due to particle attraction [Abrikosov *et al.* (1975)] (the attraction of composite fermions can be achieved by deforming the vortex [Read (1994); Greiter *et al.* (1992); Rezayi and Read (1994); Lee (1998); Willett *et al.* (1987)], but it must be added that the instability of the normal Fermi liquid in 2D may have a different character than its equivalent in 3D [Kivelson *et al.* (1986)]).

The properties of Read's vortices can be summarized as follows:

- the vorticity is defined by q in the expression (2.4) (i.e., if we use any j-th particle to describe a closed loop around the center of a vortex, a phase shift of $2\pi q$ is achieved),

---

[6] Also probable for $\nu = \frac{1}{2}$ or $\frac{1}{4}$, where the latest experimental reports [Luhman *et al.* (2008); Shabani *et al.* (2009a); Papić *et al.* (2009b); Shabani *et al.* (2009b)] suggest that incompressible states are observed, associated with the energy gap produced by pairing.

- the vortex causes depletion of the local charge, which is related to the Coulomb repulsion at small distances (higher vorticity reflects a stronger trend toward zero in the Jastrow factor, i.e., a 'reduction' of close distances); as a result, an electron charge is screened by a positive uniform background charge (*jellium*), which is locally unbalanced as a result of vortex depletion; for a vorticity q, the positive charge caused by vortex depletion is $-q\nu e$ [Read (1994)] (therefore, for $\nu = 1/q$ it is $-e$, which causes complete screening of the electron's charge; then, the electron-vortex complexes behave like neutral particles in this particular case),

- if we interchange the locations of two vortices, we produce a phase shift of $q^2\nu\pi$, (because of the charge deficit of the vortex [Read (1994)]), which is $q\pi$ when $\nu = \frac{1}{q}$ for vortices with a vorticity of q, together with the associated electrons (electrons add a charge of e to the complexes and produce an additional π phase shift upon interchange of these complexes); electron complexes with vortices behave like composite bosons with zero effective charge in the case of odd q and like composite fermions in the case of even q; these non-charged effective bosons can condense into a Bose–Einstein condensate, which reproduces the precise form of the Laughlin function for odd q [Rajaraman and Sondhi (1996)] whereas for even q we are dealing with a Fermi sea of effective discharged fermions (these are not affected by the magnetic field), which can be identified with a Hall metal [Halperin *et al.* (1993); Rezayi and Read (1994); Lee (1998)].

The main characteristic of vortices is their collective nature; the definition of a vortex includes all of the particles that together form the vortex. Therefore, a vortex is not a local structure (unlike the localized quanta of Jain's flux tubes [*flux tubes* with an idealized zero diameter]). However, vortices are useful for the interpretation of Laughlin correlations because of their convenient effect of diluting their internal charge density. This dilution is actually a manifestation of the decreasing probability density to find particles close to one another as a result of the 'pinning of multi-fold zeros' to particles by the Jastrow factor (for higher values of the p exponent in the Jastrow polynomial, $\prod_{i>j}^{N}(z_i - z_j)^p$, it goes toward zero faster upon reducing the argument).

Although Jain's and Read's approaches to composite particles are both effective, they are completely phenomenological and adjusted to the known in advance Laughlin function (in the case of Jain's composite fermions, the number of attached flux tube quanta is adjusted, and the vorticity is adjusted in the case of Read's vortices). In this way, they illustrate Laughlin correlations rather than explaining their origin. In the case of Jain's structure, flux tubes are attached to particles in order to achieve the proper phase shift during particle exchanges. When particles are modified with attached flux tubes (based on the Aharonov–Bohm effect), they have a purely model auxiliary character, and these flux tubes do not actually exist. However, the model flux tubes certainly illustrate other fact that is more real and located in the topological structure of Laughlin correlations. Therefore, identifying the topological reason for these correlations, which is the subject of this treatise, would offer an explanation for the model nature of the flux tube quanta attached to particles in the case of composite fermions.

2.2.3 *Local gauge transformations corresponding to Jain's flux tubes and Read's vortices in the structure of composite fermions*

All properties of Read's nonlocal vortices and Jain's infinitely thin flux tubes can be described by a formal local gauge transformation [Rajaraman and Sondhi (1996); Shankar and Murthy (1997)] applied to the initial fermions. If we define the initial fermions by field operators ($\Psi(\mathbf{x})$ for the annihilation of a fermion in $\mathbf{x}$ and $\Psi^+(\mathbf{x})$ for its creation), composite bosons and composite fermions can be defined as follows:

$$\Phi(\mathbf{x}) = e^{-J(\mathbf{x})}\Psi(\mathbf{x}), \quad \Theta(\mathbf{x}) = \Psi^+(\mathbf{x})e^{J(\mathbf{x})}, \tag{2.7}$$

where $J(\mathbf{x}) = q \int d^2x' \rho(\mathbf{x}')log(z - z') - \frac{|z|^2}{4l^2}$, and e^{-J} corresponds to a nonunitary gauge transformation describing the attachment of Read's vortices (or Jain's flux tubes) to initial fermions described by $\Psi(\mathbf{x})$ and $\Psi^+(\mathbf{x})$ (for the annihilation and creation operators, respectively).

If we extract the imaginary part of the expression $J(\mathbf{x})$ (i.e., only include the imaginary part of *log*), we produce the Hermitian Chern–Simons field [7]. This Hermitian Chern–Simons field exactly corresponds to the

---

[7] Chern–Simons field theory is well-developed in terms of quantum field theory; it is a chiral field [Chern and Simons (1974); Witten (1995)] in both 3D and 2D; for 2D, the Chern–Simons fields is useful for the formal and computationally efficient attachment of local flux tubes to particles and changing statistics [Fetter *et al.* (1989); Lopez and Fradkin (1991)], and it has been widely used for transmutation (change) of 2D statistics

assignment of local flux tubes to fermions [Zhang *et al.* (1989)]. Regarding the field operators $\Phi(\mathbf{x})$ and $\Theta(\mathbf{x})$, $\Phi^+(\mathbf{x}) = \Theta(\mathbf{x})e^{-J(\mathbf{x})-J^+(\mathbf{x})}$, although they are not mutually conjugated in the general case, they are perfectly mutual conjugated in the case of the Hermitian Chern–Simons field, and they generally describe composite bosons (for odd q) and composite fermions (for even q), which can easily be verified with a commutator [Rajaraman and Sondhi (1996)]. Within the mean field approximation [Rajaraman and Sondhi (1996)], it is crucial to note that the real part of J in this approximation disappears because the real part of log is reduced by the Gaussian, while the Hermitian Chern–Simons field is reduced by the external magnetic field. By virtue of the relation $e^{q\sum_j log(z-z_j)} = \prod_j^N (z - z_j)^q$ (for the density operator $\rho(\mathbf{x}) = \Psi^+(\mathbf{x})\Psi(\mathbf{x}) \implies \sum_{j=1}^N \delta(z - z_j)$), one can expect that the equation given by (2.7) would reproduce all properties of the vortices, which corresponds to Read's vortex definition. Nevertheless, this definition does not explain the causes of any particular vorticity. Instead, it offers an elegant form that is convenient for calculation purposes.

The transformation (2.7) and the related vortex image enable the interpretation of Laughlin states as Bose–Einstein condensates of composite bosons in the case of LLL fillings, $\nu = \frac{1}{q}$, q – odd, [Read (1994); Rajaraman and Sondhi (1996); Zhang *et al.* (1989)], and it enables the interpretation of the system as a compressible (not blocked with a gap) Fermi sea for q – even [Pasquier and Haldane (1998); Rezayi and Read (1994); Lee (1998)] (the latter is unstable for BCS coupling) [Greiter *et al.* (1992); Moore and Read (1991); Abrikosov *et al.* (1975)], in full accordance with the previously presented modeling of Laughlin correlations through Read's vortices.

The specific character of the vortices is especially clearly visible in the case of $\nu = 1$. The vortices (2.3) attached to electrons for $\nu = 1$ lead to the Vandermonde factor (i.e., the Jastrow polynomial with $q = 1$). In this case, the proper Laughlin state takes the form of the Slater function for N *non-interacting* fermions, which can be described with equal effectiveness by a condensate of composite bosons [Read (1998)] defined with the use of the transformation (2.7) for $q = 1$ (interestingly, the entire impact of the magnetic field on the fermions is replaced by a Bose–Einstein condensate of field-resistant uncharged bosons). The Coulomb interaction does not contribute here, as in the case when $\nu = 1$ the Haldane pseudopotentials [Prange and Girvin (1990); Haldane (1983)] do not contribute (i.e., the

---

and modeling of anyons and composite fermions [Heinonen (1998); Lopez and Fradkin (1991)].

significant short-range part of the Coulomb interaction that determines the form of the Laughlin function equals zero, because $q - 2 < 0$, for $q = 1$); thus, the Slater function for *non-interacting* particles is appropriate as the ground state of the Hamiltonian for interacting particles when $\nu = 1$.

Chapter 3

Topological methods for the description of many particle systems at various manifolds

3.1 Braid groups

The tool used to include the topological characteristics of systems of many particles in spaces with various dimensions and located on different manifolds (including compact manifolds and not-simply-connected spaces) is algebraic topology [Spanier (1966)], which reflects the complexity of the trajectories of many particle systems in terms of homotopy (expressing the topological properties of continuous trajectory deformations). The fundamental group for a given space D or the first homotopy group of this space (cf. Chapter 5.4), marked with $\pi_1(D)$, is a collection of topologically non-equivalent (nonhomotopic) classes of closed trajectories in the space D. If this space is a configuration space of a system of N identical particles, each of which traverses its own trajectory on the M manifold, the appropriate π_1 homotopy group is called a braid group. The configuration space for N identical particles located on the M manifold (e.g., R^n or a compact manifold, such as a sphere or a torus) is defined as follows: $Q_N(M) = (M^N \setminus \Delta)/S_N$, for indistinguishable identical particles and $F_N(M) = M^N \setminus \Delta$, for distinguishable identical particles; M^N is an N-fold Cartesian product of the M manifold, Δ is a set of diagonal points (for which the coordinates of two or more particles coincide) that must be subtracted in order to assure the conservation of the number of particles in the system, and S_N is a permutation group (the quotient structure has been introduced in order to include the indistinguishability of quantum particles). The indistinguishability of particles has been artificially introduced by the definition of the configuration space, which suggests that this property is independent of quantum uncertainty principles.

The braid group is the first homotopy group [Spanier (1966); Mermin (1979)], π_1, of the configuration space of a system of N particles. The elements of $\pi_1(A)$ are the topologically non-equivalent classes of closed trajectories in A space. In the case of a system of N particles, when we consider braid groups, A is the N-particle configuration space. Braid groups represent the only possible classical trajectories (closed) of a system of N particles (while not referring to any specific dynamics), whereas quantization is carried out with the unitary representations of classical braid trajectories as part of the Feynman formalism of integrals over trajectories, which is summarized in the subsequent chapters.

Two types of braid groups are defined for two different configuration spaces for indistinguishable and distinguishable identical particles, as presented below [Birman (1974)]. The full braid group ,

$$\pi_1(Q_N(M)) = \pi_1(M^N \setminus \Delta)/S_N, \tag{3.1}$$

and the pure braid group ,

$$\pi_1(F_N(M)) = \pi_1(M^N \setminus \Delta). \tag{3.2}$$

3.1.1   Full braid groups for R^3, R^2, sphere S^2 and torus T

For $M = R^n$ and $n > 2$, the braid groups have a simple structure. The full braid groups for $n > 2$ are equal to the S_N permutation groups (the permutation groups S_N are finite and of $N!$ order). For $M = R^2$ (as well as for compact, locally two-dimensional manifolds such as a sphere or a torus in three dimensions), braid groups are infinite groups with highly complex structures.

For the sake of convenience and direct insight, the structure of the braid groups for the (R^2) plane can be presented via simple geometrical graphs by using the so-called geometric braids [Birman (1974); Jacak *et al.* (2003)], as shown in Fig. 3.1. This figure presents the following: a) the geometric braid corresponding to the σ_i generator of the full braid group (i.e., to the interchange of the i-th particle with the $(i+1)$-th one, where the lines represent particle trajectories), b) the geometric braid corresponding to the reversed generator element, σ_i^{-1}, and c) the geometric braid of the square of the generator $(\sigma_i)^2 \neq e$ (e represents the neutral element of the group). In three dimensions, the equation $(\sigma_i)^2 = e$ holds, whereas in two dimensions, $(\sigma_i)^2 \neq e$, which is the reason for the complexity (of infinite type) of the braid structure for two-dimensional manifolds.

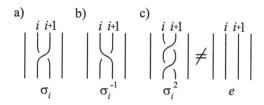

Fig. 3.1   Geometric presentation for B_N (traditional [Artin (1947)] marking for a full braid group for the R^2 manifold): σ_i generator (a) and its reverse element, σ_i^{-1} (b); square of generator σ_i^2 (c).

One can list the formal conditions for generators σ_i, $i = 1, ..., N - 1$, in order to completely define the full braid group in an abstract manner [Artin (1947); Birman (1974)]. These conditions have the form shown below, and their respective geometric braids are presented in Fig. 3.2, (a,b):

$$\sigma_i \sigma_{i+1} \sigma_i = \sigma_{i+1} \sigma_i \sigma_{i+1}, \text{ for } 1 \leq i \leq N - 2, \tag{3.3}$$

$$\sigma_i \sigma_j = \sigma_j \sigma_i. \text{ for } 1 \leq i, j \leq N - 1, |i - j| \geq 2. \tag{3.4}$$

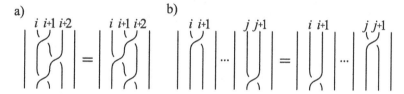

Fig. 3.2   Formal conditions for generators defining the full braid group for R^2 (*a* represents equation (3.3) and *b* represents equation (3.4)).

In the case of a S^2 sphere, there is additional condition (apart from the identical conditions for B_N group) for generators [Birman (1974)],

$$\sigma_1 \cdot \sigma_2 \cdot ... \cdot \sigma_{n-2} \cdot \sigma_{n-1}^2 \cdot \sigma_{n-2} \cdot ... \cdot \sigma_2 \cdot \sigma_1 = e, \tag{3.5}$$

which reflects the fact that the loop encircled by a given particle on a sphere around all the other particles is homotopic with a point. For the T torus, additional relations [Einarsson (1990)] correspond to two non-equivalent trajectories that are available to each particle on this not-simply-connected manifold [Jacak *et al.* (2003); Birman (1969)].

3.1.2   *Pure braid group*

The preservation of the initial ordering of particles is not required in the case of braids from the full braid group (the particles are indistinguishable;

thus, their ordering is the same up to an arbitrary permutation), while the initial ordering of particles must be unchanged for braids from the pure braid group (in which the particles are distinguishable). l_{ij} generators of pure braid group (Fig. 3.3) [Birman (1974)] correspond to the exchange of particle pairs, i, j, with no permutation of the particle ordering, and they can be presented with the use of generators σ_i,

$$l_{ij} = \sigma_{j-1} \cdot \sigma_{j-2} \cdots \sigma_{i+1} \cdot \sigma_i^2 \cdot \sigma_{i+1}^{-1} \cdots \sigma_{j-2}^{-1} \cdot \sigma_{j-1}^{-1}, \quad 1 \geq i \geq j \geq N - 1. \quad (3.6)$$

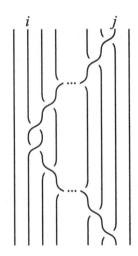

Fig. 3.3   Generator l_{ij} of the pure braid group.

Therefore, a pure braid group is a subgroup of the full group because l_{ij} generators can be expressed by σ_i generators. Pure braid group generators are defined by the following relations [Birman (1974); Jacak *et al.* (2003)]:

$$l_{rs}^{-1} \cdot l_{ij} \cdot l_{rs} = \begin{cases} l_{ij}, & i < r < s < j \\ l_{ij}, & r < s < i < j \\ l_{rj} \cdot l_{ij} \cdot l_{rj}^{-1}, & r < i = s < j \\ l_{rj} \cdot l_{sj} \cdot l_{ij} \cdot l_{sj}^{-1} \cdot l_{rj}^{-1}, & i = r < s < j \\ l_{rj} \cdot l_{sj} \cdot l_{rj}^{-1} \cdot l_{sj}^{-1} \cdot l_{ij} \cdot l_{sj} \cdot l_{rj} \cdot l_{sj}^{-1} \cdot l_{rj}^{-1}, & r < i < s < j. \end{cases}$$
$$(3.7)$$

It is worth noting the fact that the relationship between the full braid group and the pure braid group is expressed by a quotient structure [Birman (1974)], $B_N / \pi_1(F_N(R^2)) = S_N$ (B_N is a marking for the full braid group for a plane) [Artin (1947)].

3.2 Feynman integrals over trajectories and the relation with the one-dimensional unitary representations of the full braid group

The quantization of a system of N indistinguishable identical particles can be carried out with the use of the Feynman path integration formalism [Feynman and Hibbs (1964); Chaichian and Demichev (2001a,b); Papadopoulos and Devreese (1978)]. By using this formalism, we can obtain the expression for the propagator (the matrix element of the evolution operator in position representation that gives probability for the system to transition from point a at time t_1 to point b at the time t_2 in the configuration space) in the following form:

$$I_{a \to b} = \int d\lambda e^{iS[\lambda_{a,b}]/\hbar}, \qquad (3.8)$$

where $S[\lambda_{a,b}]$ is a classical action for the $\lambda_{a,b}$ trajectory in the classical configuration space for a system of N particles and $d\lambda$ is a measure in the trajectory space. For every trajectory joining points a and b in the configuration space of a system with N particles, one can attach additional closed loops that are elements of the full braid group. The trajectories produced by this link fall into topologically separate classes represented by the elements of the full braid group (which contains closed trajectories). Because all possible trajectories should be included in the path integral, including those with loops of various dimension and location along the path, the entire trajectory family decomposes into the sum of disjoint (in the sense of continuous transformations) trajectory subsets, which can be enumerated with the full braid group elements. Continuity conditions allow the measure in the trajectory space to be defined only within each such separate subset, but all the nonhomotopic trajectory subsets must be taken into account. Therefore, an additional unitary factor (an unitary weight corresponding to the given class of trajectories) must be introduced to the formula for integration over the trajectories [Laidlaw and DeWitt (1971); Wilczek (1990); Wu (1984)], as well as the additional summation over the elements of the braid group (because every element of the braid group can be attached to any loopless open trajectory $\lambda_{a,b}$):

$$I_{a \to b} = \sum_{l \in \pi_1} e^{i\alpha_l} \int d\lambda_l e^{iS[\lambda_{l(a,b)}]/\hbar}, \qquad (3.9)$$

where π_1 stands for the full braid group. The $e^{i\alpha_l}$ factors form a one-dimensional unitary representation (1DUR) of the full braid group [Laidlaw and DeWitt (1971)]. Different representations correspond to different types of quantum particles related to the same classical particles. During the integration over trajectories, all possible trajectories contribute, regardless of the presence of fields or other physical conditions. However, this contribution only occurs when these trajectories are not excluded, as it happens in the case of strong magnetic field presence for two-dimensional charged particles, which will be discussed below.

Classical trajectories from the full braid group have no quantum interpretation. Quantum particles do not traverse any braid trajectories because these particles have no trajectories at all in the quantum sense. However, according to the general quantization rules [Imbo and Sudarshan (1988); Sudarshan *et al.* (1988)], a N-particle wave function must transform in accordance with the one-dimensional unitary representation of a particular element of a braid group if classical particles traverse along a closed trajectory in the configuration space of the N-particle system corresponding to that particular braid element (this describes the exchange of arguments of a N-particle wave function, which cannot be reduced to a simple permutation in the 2D case). Because braids from the full braid group describe exchanges of particles, the corresponding one-dimensional unitary representations are phase factors that determine their statistics.

3.3   Bosons, fermions, anyons and composite particles

3.3.1   *Anyons on the plane, sphere and torus*

For S_N, which is a full braid group for manifolds of three dimensions and of higher dimensions , there are only two different one-dimensional unitary representations (1DURs),

$$\sigma_i \to \begin{cases} e^{i0}, \\ e^{i\pi}, \end{cases} \tag{3.10}$$

corresponding to bosons and fermions, respectively (which are expressed by specific symmetrical and anti-symmetrical wave functions because the braid group in this case is a simple permutation group). For a two-dimensional space (plane), the braid group (structurally more complex than S_N) has an infinite number of one-dimensional unitary representations [Jacak *et al.* (2003); Imbo *et al.* (1990)], which are recorded for generators of the group as $\sigma_i \to e^{i\theta}$, $\theta \in (-\pi, \pi]$, where the values of θ enumerate various types of so-called anyons [Laughlin (1983b,a); Das Sarma and Pinczuk (1997);

Wilczek (1990); Wu (1984); Nayak *et al.* (2008)]. The elements of the one-dimensional unitary representations of a full braid group do not depend on the i index (of the σ_i generator) because of the condition imposed on the generators. Because the 1DUR elements commute, we can write (using the conditions given in (3.3) for generators of the full braid group B_N) $e^{i\theta_i} = e^{i\theta_{i+1}}$, where $\sigma_i \to e^{i\theta_i}$, which gives us independence from the index i of the 1DUR elements.

For a sphere S^2, the one-dimensional unitary representations have the form [Imbo *et al.* (1990)], $e^{i\theta}$, where $\theta = k\pi/(N-1)$, $k = 0, 1, 2, ..., 2N-3$. Interestingly, for two particles on a sphere (i.e., for $N = 2$, $k = 0, 1$), only bosonic or fermionic statistics are available (this is a direct consequence of the condition (3.5)), and anyons may appear on the sphere only after there have been at least 3 particles. In the case of a T torus, for any number of N particles, only $\theta = 0$ or π are allowed [Imbo *et al.* (1990); Einarsson (1990)]; thus, anyons *do not exist* on a torus, apart from fermions and bosons. The last result has been extended [Imbo *et al.* (1990)] for all compact locally two-dimensional manifolds, with the exception of the sphere.

3.3.2 *Quantum statistics and braid groups*

As has been mentioned earlier, unitary representations (particularly one-dimensional unitary representations of the full braid group) serve to identify various types of quantum particles, which correspond to the same type of classical particles [Wilczek (1990); Wu (1984)].

It is worth to highlight once again that for the S_N permutation group, which is the full braid group for R^n, $n \geq 3$, there are only two one-dimensional unitary representations: $\sigma_i \to e^{i\pi}$ and $\sigma_i \to e^{i0}$ (σ_i represents the exchange of particles i-th and $(i+1)$-th), corresponding to fermions and bosons. For R^2, because of the significantly more complex structure of the braid group, we deal with an infinite number of distinct one-dimensional unitary representations that correspond to the infinite number of different quantum realizations of the same classical model for anyons (abelian anyons because the one-dimensional unitary representations commute).

It is important to note that the 2π periodicity of the factor $e^{i\theta}$, $\theta \in (-\pi, \pi]$, prevents the statistical distinction of composite fermions from ordinary fermions with the use of one-dimensional unitary representations of the full braid group (as composite fermions need $\Theta = 3\pi, 5\pi, ...$, while ordinary fermions are associated with $\Theta = \pi$). To overcome this ostensible obstacle, we propose that composite fermions can be related to one-dimensional unitary representations of appropriately constructed subgroups of braid groups, instead of the full braid group, thus introducing the opportunity to distinguish composite fermions from ordinary fermions.

Anyons can be considered to be 'two-dimensional' quantum particles only if there is no strong magnetic field because the related braid trajectories are selected from the full braid group without taking into account any modifications caused by the cyclotron movement. Nevertheless, the presence of a magnetic field has a major impact on trajectories, the resulting classical cyclotron movement limits the diversity of the available trajectories, especially in the case in which the distance between the particles is greater than double the cyclotron radius, which prevents the exchange of particles along cyclotron trajectories and forces (as shown below) particle exchanges along the trajectories with additional loops. Thus, in the case of a strong magnetic field in two dimensions, not all trajectories are possible, and the ones that are impossible should be removed from the domain of the integral over the trajectories. Thus the presence of the magnetic field necessitates the construction of trajectories from fragments of cyclotron trajectories, which limits the number of available trajectories (regardless of the specific dynamics details), especially when the cyclotron radii become too short in comparison to the distance between the particles. This situation occurs for fractional fillings of the LLL (let us note that the distance between the particles is stabilized by the short-range part of the Coulomb repulsion at the level resulting from the planar density of particles). In a special case of fractional filling $\frac{1}{p}$ of the LLL, cyclotron radii that are too short require trajectories to be assigned to appropriate subgroups of the braid group that are made of trajectories with additional loops (they will remain if one removes not attainable single-loop braids). This requirement increases the cyclotron radius and re-enables particle exchanges, as will be illustrated and explained below. In the presence of a magnetic field, the sum in the Feynman propagator is limited to the elements of this subgroup (actually to a semi-group for a specific direction of the magnetic field); i.e., according to the restrictions of cyclotron movement, the propagator is limited to selected trajectory classes instead of all elements of the full braid group.

We can conclude that one-dimensional unitary representations of a subgroup of the braid group that are generated by cyclotron movement in strong magnetic fields determine the statistics of composite fermions. Consequently, we can implement composite fermions as distinct two-dimensional quantum particles corresponding to the appropriately selected one-dimensional unitary representations of the cyclotron braid subgroup. Composite fermions can then be assigned to one-dimensional unitary representations of cyclotron subgroups of the full braid group (i.e., to a separate braid object), which enables the distinction between composite and ordinary fermions, as well as other 2D quantum particles.

In this way we can avoid the need to use an artificial models for composite fermions as a structure consisting of fictitious auxiliary localized flux tubes of the magnetic field attached to fermions. In this novel approach using the cyclotron braid subgroup, an additional phase of the wave function is generated by an additional loop during the particle exchange as an inevitable property caused by the presence of a magnetic field in a two-dimensional system of charged particles with fractional filling $\nu = \frac{1}{p}$ of the Landau level.

3.4 Multidimensional unitary irreducible representations of braid groups

An important role is also played by multi-dimensional unitary irreducible representations (MDURs) of braid groups. According to Kitaev's concept [Nayak *et al.* (2008); Kitaev (2003)], any unitary evolution of a multi-qubit system (e.g., a two-qubit quantum gate for QIP [Quantum Information Processing]) may be approximated by multi-dimensional unitary irreducible representations (of the appropriate order) of the full braid group, provided that the multi-dimensional unitary representations of the braid group in a space of unitary matrices have a sufficient density [Nayak *et al.* (2008)]. The MDURs can be associated with degenerate low-energy excitations (quasiparticles or quasiholes, which are usually treated as anyons[1]) beyond the ground state for the particular fractional filling of the Landau level. The elements of MDURs do not commute with one another because they are matrices; consequently, these degenerate anyon states of quasiparticle excitations are referred as non-abelian anyons [Nayak *et al.* (2008)]. Unfortunately, non-abelian anyons (which have recently been studied, particularly for low-energy excitations at Landau level filling factors of $\frac{5}{2}$ and $\frac{12}{5}$) most likely correspond to insufficiently dense MDURs (for non-abelian anyons in the case of $\frac{5}{2}$, MDURs are not dense enough to reach the appropriate

---

[1]Identification of the anyon character of quasiparticles can be be carried out by means of numerical calculation of the Berry phase on a closed trajectory for a wave function, which models the quasiparticle or quasihole.

approximation of multi-qubit gates [Nayak *et al.* (2008)], whereas the second currently discussed state for $\frac{12}{5}$ remains controversial [Xia *et al.* (2004)]). This effort increases the significance of the search for other possible systems with fractional statistics that have dense MDURs related to non-abelian anyons. In the following chapters, we will introduce cyclotron braid subgroups of the full braid group. Because the subgroups usually have richer representations than the group, we can expect that the cyclotron braid subgroups will prove to be useful for expanding the topological methods of the QIP because the appropriate MDURs of the cyclotron subgroup can be denser than the representations of the full braid group.

Chapter 4

Cyclotron braids for multi-particle-charged 2D systems in a strong magnetic field

4.1 Insufficient length of cyclotron radii in 2D systems in a strong magnetic field

One-dimensional unitary representations of the full braid group [Wu (1984); Birman (1974); Jacak *et al.* (2003); Imbo *et al.* (1990)], which is the homotopy group π_1 of the configuration space for N indistinguishable identical particles [Birman (1974)], define the weights for integrals over trajectories [Wilczek (1990); Wu (1984); Laidlaw and DeWitt (1971)]. If the trajectories belong to separate homotopy classes, then defining a measure in the whole space of the trajectories is not possible because of discontinuity. The measure can only be restricted to separate homotopy classes of trajectories; however, we should additionally sum over all topologically non-equivalent classes with unitary weight factors. Non-equivalent trajectory classes result by attaching to open trajectories $\lambda_{a,b}$ (joining points a and b in the configuration space) additional closed multi-particle loops; the number of nonequivalent loops is the same as the number of elements in the full braid group, and they are all mutually nonhomotopic. Therefore, the Feynman integral over the trajectories, which determines the quantum probability of system passing between $a, t = 0$ and b, t (Feynman propagator) [1], takes the following form [Laidlaw and DeWitt (1971)],

$$I_{a \to b} = \sum_{l \in \pi_1} e^{i\alpha_l} \int d\lambda_l e^{iS[\lambda_{l(a,b)}]/\hbar}, \qquad (4.1)$$

where π_1 is the appropriate braid group. The weight factors, $e^{i\alpha_l}$, constitute one-dimensional representations (1DURs) of the full braid group,

---

[1] It is a matrix element in the position representation of the evolution operator [Chaichian and Demichev (2001a,b); Feynman and Hibbs (1964)].

and different representations correspond to different types of quantum particles. For the permutation group S_N, which is a full braid group of N particles in R^n where $n \geq 3$, there are only two one-dimensional unitary representations: $\sigma_i \to e^{i\pi}$ and $\sigma_i \to e^{i0}$, (σ_i stands for the exchange of the i-th and $(i+1)$-th particles, i.e., the group generators), which correspond to bosons and fermions, respectively. For N particles in R^2, the braid group is considerably more expanded than S_N and has an infinite number of one-dimensional unitary representations [Wu (1984); Imbo *et al.* (1990)] that are defined for generators of the braid group as $\sigma_i \to e^{i\theta}$ where $i = 1, ..., N-1$, $\theta \in (-\pi, \pi]$ and various values of θ correspond to various types of anyons [Laughlin (1983a); Wilczek (1990); Wu (1984)] (abelian anyons because the elements of the one-dimensional representations commute).

Closed trajectories from the full braid group describe the exchanges of identical particles; therefore, their one-dimensional unitary representations (1DURs) determine the particle statistics. Because these representations (1DURs) are periodic phase factors that have a period of 2π, distinguishing between the 1DURs of composite fermions (associated with the Laughlin correlations) from ordinary fermions was not possible because composite fermions require phase shifts of $p\pi$, $p = 3, 5...$ and $e^{ip\pi} = e^{i\pi} = -1$ (p–odd). If the association of composite fermions with the one-dimensional unitary representations (1DURs) of the full braid group is not possible, we suggest [Jacak *et al.* (2009a)] associating the composite fermions with properly defined *subgroups* of the full braid group, thereby making a distinction between composite and ordinary fermions.

Full braid groups include all of the possible closed multi-particle classical trajectories in the form of braids (with a possible permutation shift of the initial and final ordering of particles because of their indistinguishability). Our concept is based on the observation that the inclusion of the magnetic field in the systems of 2D charged particles considerably changes these trajectories and that the classical cyclotron movement substantially limits the diversity of possible braids.

For cases in which the distance between particles is greater than the double cyclotron radius, which occurs in the case of LLL fractional fillings, the exchange of particles along normal single-loop cyclotron trajectories is impossible, as the cyclotron orbits are then *too short*. However, particle exchanges are crucial for determining statistics; therefore, the cyclotron orbits must somehow increase to again enable particle exchanges. An increase in the size of the cyclotron trajectories can be achieved by reducing the effective strength of the magnetic field or by reducing the effective particle charge. These two possibilities lead to two phenomenological concepts of composite fermions, where the reduced resultant magnetic field is screened by the averaged local fictitious Jain's flux tubes [Jain (1989)] or the charge is screened in terms of the Read's vortices construction due to vortex charge depletion [Read (1994)]. Both model structures for composite fermions seem to have nothing in common with braid groups, but in fact, both of these phenomenological tricks correspond to a more natural and fundamental concept of limiting braids by excluding unattainable trajectories [Jacak *et al.* (2009a, 2010b)], which leaves us with only the available trajectories. We therefore argue that for a strong magnetic field in a 2D system of charged N particles, multi-loop braids allow the required increase in the size of cyclotron orbits that restores the particle exchanges in a natural way [Jacak *et al.* (2010b)]. These multi-loop braids form a subgroup of the full braid group, and in the presence of a strong magnetic field, the summation in the Feynman propagator must be reduced to this subgroup [Jacak *et al.* (2010a, 2009a)], along with the remaining available nonhomotopic trajectory classes obtained after excluding the impossible ones, i.e., after excluding the unavailable trajectories that have too short cyclotron radii [2].

4.2 Definition of the cyclotron braid subgroup and its unitary representations

We suggest referring composite anyons (including composite fermions) to one-dimensional unitary representations of *cyclotron subgroups* of the full braid group, which are generated by the following generators [Jacak *et al.* (2009a)]:

$$b_i^{(p)} = \sigma_i^p, \quad (p = 3, 5, 7, 9....), \quad i = 1, ..., N-1, \qquad (4.2)$$

where every p corresponds to subsequent cyclotron subgroup of type p (and hence to other type of composite particles that are associated with its representations), and σ_i, $i - 1, ..., N$ are the generators of the initial full braid group. The group element $b_i^{(p)}$ represents the new elementary exchange of i-th particle with the $(i + 1)$-th particle, with $\frac{p-1}{2}$ loops, which is clear by

---

[2]Actually, the available trajectories are assigned to a semi-group of this subgroup for the chosen orientation of the magnetic field; this semi-group has the same 1DURs as the entire subgroup.

virtue of the definition of the previous elementary exchange, σ_i (as presented in Fig. 4.2). The generators, $b_i^{(p)}$, build up the subgroup of the full braid group because they are expressed by the generators, σ_i, of the full group. However, the $b_i^{(p)}$ generators do not satisfy the condition (3.3) for generators of B_N, i.e., $b_i^{(p)} b_{i+1}^{(p)} b_i^{(p)} \neq b_{i+1}^{(p)} b_i^{(p)} b_{i+1}^{(p)}$, while the condition (3.4) is in force for $b_i^{(p)}$: $b_i^{(p)} b_j^{(p)} = b_j^{(p)} b_i^{(p)}$, for $1 \leq i, j \leq N-1$, $|i-j| \geq 2$ (Fig. 4.1).

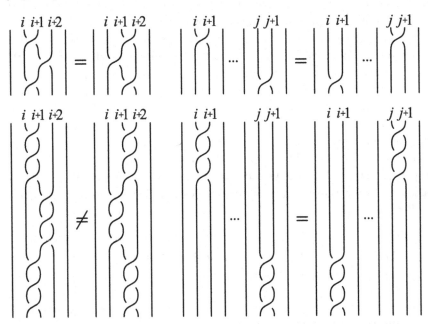

Fig. 4.1 The conditions imposed on the generators of the cyclotron braid subgroup: the top and bottom figures present the relations imposed on generators from the B_N group and for cyclotron subgroup ($p = 3$), respectively; only the second condition is fulfilled for cyclotron subgroup.

Because the condition (3.3) is not fulfilled for the cyclotron subgroup, its 1DURs may generally depend on the i index. However, unitary 1DURs of the full braid group that are reduced to the cyclotron subgroup do not depend on the i index, and the induced representations of the cyclotron subgroup of type p have the following form:

$$b_i^{(p)} \to e^{ip\alpha}, \ i = 1, ..., N-1, \tag{4.3}$$

where p is an odd integer and $\alpha \in (-\pi, \pi]$. These 1DURs, which are enumerated by *pairs* (p, α), describe composite anyons (in particular, composite fermions in the case of $\alpha = \pi$). Therefore, to distinguish between the various types of composite particles, we need to distinguish the two-parameter markings (p, α) of the 1DUR representations of cyclotron braid subgroups, which are numbered by p. We can add here that for a fixed orientation of the magnetic field, the cyclotron trajectories would rather form a semi-group of the cyclotron subgroup but with the same representations as the entire subgroup.

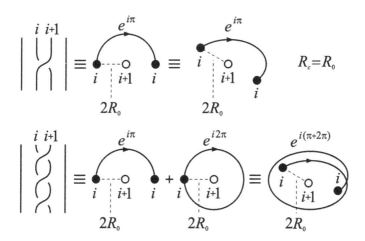

Fig. 4.2 σ_i generator for the full braid group and the corresponding relative trajectory for the exchange of i=th and $(i + 1)$-th particles (top figure); cyclotron subgroup generator, $b_i^{(p)} = \sigma_i^p$ (on figure, $p = 3$), corresponds to additional $\frac{p-1}{2}$ loops when the i-th particle exchanges with the $(i + 1)$-th particle (additional loops result in an additional phase shift of 2π each; $2R_0$ represents the distance between particles) (bottom figure).

In the presented structure, a N-particle wave function gains the proper phase shift associated with a specific multi-loop braid corresponding to the particle exchange in the language of braid groups. In this way, the Aharonov–Bohm phase shift, which results from auxiliary flux tubes that are localized on particles in Jain's model of composite fermions, is replaced with additional loops in the braids (each loop adds 2π to the phase shift, which is illustrated in Fig. 4.2 (right)). Emphasis must be placed on the fact that quantum particles do not traverse braid trajectories, as quantum particles do not have trajectories at all. Nevertheless, according to general quantization rules [Imbo and Sudarshan (1988); Sudarshan *et al.* (1988)], the N-particle wave function must transform according to a one-dimensional unitary representation (1DUR) of some element of the braid group where the particles traverse, in classical terms, a closed loop corresponding to this particular braid element (which represents the arguments exchange in a wave function). Because braids describe the exchanges in the location of particles in a classical sense (as the arguments exchanges in a multiparticle wave function), their 1DURs represent the possible quantum statistics of the system. Therefore, for the cyclotron braid groups of the trajectories that are generated by the generators $b_i^{(p)}$, $i = 1, ..., N - 1$, we obtain a statistical phase shift, $p\pi$, for composite fermions (i.e., for $\alpha = \pi$ in equation (4.3)), as required by the Laughlin correlations, without the need to refer to models using auxiliary flux tubes.

4.3   Multi-loop trajectories—the response of the system to cyclotron trajectories that are too short

In line with the content of the previous section, we have suggested associating the composite particles and Laughlin correlations with the 1DURs *of cyclotron braid subgroups*, which are generated by the generators: $b_i^{(p)} = \sigma_i^p$, $(p = 3, 5...)$, $i = 1, ..., N - 1$, with different values of p for different types of composite particles (σ_i are the generators of the full braid group). $b_i^{(p)}$ generators represent the exchanges of the i-th and $(i + 1)$-th particles, with $\frac{p-1}{2}$ loops (there are no exchanges with smaller numbers of loops in this subgroup), which is visible in the presentation by σ_i (Fig. 4.2). 1DURs of the full braid group cut to subgroups (they do not depend on i then) describe 1DURs of the subgroup: that $b_i^{(p)} \rightarrow e^{ip\alpha}$, $i = 1, ..., N - 1$, where p is an odd integer, $\alpha \in (-\pi, \pi]$; these representations are numbered by pairs (p, α) and describe composite anyons (or composite fermions when $\alpha = \pi$).

A N-particle wave function aquires the phase shift required for Laughlin correlations in particle exchanges because, according to quantization principles [Sudarshan *et al.* (1988); Imbo *et al.* (1990)], a wave function gains the phase shift of a braid 1DUR when the particles exchange according to this particular braid element. Each additional loop in the braid adds 2π to the phase shift if we assume $\alpha = \pi$, which is suitable for composite fermions, as shown in Fig. 4.2 (lower).

It should be noted that, although quantum particles do not move along the braids, the arguments exchanges of a N-particle wave function correspond to appropriate braids in 2D, *not to simple permutations* as was the case in 3D. Therefore, for the cyclotron braids $b_i^{(p)}$, $i = 1, ..., N - 1$, we obtain a phase shift, $p\pi$, that is expected for composite fermions (i.e., for $\alpha = \pi$), which reproduces the phase required by Laughlin correlations without the need to introduce local flux tubes or vortices.

Every additional loop in the relative trajectory of particle exchange, which are defined by a generator $b_i^{(p)}$ of the cyclotron subgroup, reproduces an additional loop of the individual cyclotron trajectories of both exchanging particles, which form the relative trajectory of the exchange, as depicted in figure 4.3. Individual cyclotron trajectories of two interchanging particles are repeated by the relative trajectory (Fig. 4.3 (c,d) with a radius twice as long as that of cyclotron trajectories of individual particles (Fig. 4.3 (a,b). In quantum language, referring to classical multi-loop cyclotron trajectories, one can only estimate the number of external field flux quanta that are assigned to each particle in the system, $\frac{BS}{N} / \frac{hc}{e}$, which for the filling fraction $\frac{1}{p}$ of LLL is equal to p and is the same as the number of cyclotron loops. Therefore, we can formulate a simple rule: (*in the case of filling* $\frac{1}{p}$ of LLL [p odd]): an additional loop on a cyclotron braid that describes the particle exchange leads to <u>two</u> additional quanta of the magnetic field flux that passes through the individual cyclotron trajectory of every particle.

The above rule directly results from the definition of the cyclotron trajectory, which must be a closed individual trajectory of a particle and hence, must correspond to a <u>double</u> exchange of a pair of particles. The individual trajectories of both exchanged particles will then be closed, which replicates the closed relative trajectory of the double exchange (a single exchange has an open trajectory).

If the exchange is simple, i.e., with no additional loops, then the corresponding individual cyclotron trajectories of the particles are also simple (single-loop). In the case when the exchange takes place with additional loops, as in the cyclotron p ($p > 1$) subgroup, the trajectory of the double exchange (closed) has $2\frac{p-1}{2} + 1 = p$ full loops, and the individual cyclotron trajectories of particles have just as many, i.e., p full loops [Jacak *et al.* (2010b, 2011)].

It is very important to highlight the difference between coils (e.g., of wire in a 3D coil) and multi-loop 2D cyclotron trajectories. In the latter case, additional 2D loops *cannot* increase the total magnetic field flux that passes through the system, and all of the multiple loops must share together this total flux. 3D coils are different in that every new coil adds a new portion to the flux, as every coil adds its area, through which the magnetic field is passing, thereby increasing the total flux (this is not the case in 2D).

Therefore, we can see that in 2D systems, additional cyclotron loops take away fragments of the total flux per particle (in the case of fillings $\frac{1}{p}$, this fragment is the flux quantum) and reduce the magnetic field; this explains the nature of the Jain's auxiliary flux tubes, which were attached to composite fermions in order to screen the external field B. Therefore, composite fermions are not composite structures made of particles and attached flux tubes, although we preserve their original name. Composite fermions are rightful, as e.g., fermions, different in statistics and separate particles that are related to cyclotron subgroups (separate topological objects) and their representations. Similarly, a somewhat confusing name, which refers again to the traditional model structures, can be used in the case of composite anyons that are associated with the fractional representations 1DURs of cyclotron subgroups (i.e., with a fractional α in the representation (4.3)). Composite anyons are also not composite objects; they are rightful, separated in statistics terms particles in 2D in the presence of strong magnetic fields, which are similar to ordinary anyons in 2D in the absence of a magnetic field.

Therefore, Jain's theory of composite fermions with auxiliary flux tubes [Jain (1989); Heinonen (1998); Jain (2007)], as well as Read's formulation with vortices [Read (1994); Greiter *et al.* (1992); Pasquier and Haldane (1998); Rajaraman and Sondhi (1996); Shankar and Murthy (1997)] attached to particles, are only effective, heuristic formulations with an *a posteriori* model character with fictitious elements introduced in order to mimic the Laughlin function properties that are known beforehand. In Jain's construction, a specific number of quanta flux tubes are used to reproduce the Laughlin phase shift when the particles exchange, whereas the vorticity of Read's vortices corresponds to the exponent in the Laughlin function (the vortices themselves have the form of factors in Jastrow polynomial and reflect the 'pinning' of multi-fold zeros to particles, due to the form of the Jastrow polynomial). However, in the braid formulation, we are looking for more fundamental arguments that would explain a particular Laughlin phase shift when particles exchange. We identify this fundamental cause as the cyclotron radii not being long enough to enable exchanges in fields corresponding to fractional fillings of LLL, which, in a natural and consistent way leads to multi-loop braid trajectories, along which particle exchanges are still possible. Multi-loop braid trajectories share the same common total flux and thus reduce effective field strength in 2D systems because they do not change the total area of the planar system and must therefore have a larger radius with the same flux passing through the multi-looped cyclotron trajectory, as each loop is then pierced by a single flux quantum (for fillings $\frac{1}{p}$) – which is enough to reach other particles to carry out the particle exchange.

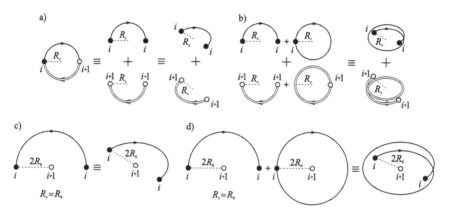

Fig. 4.3 Half of the cyclotron trajectory of the individual particles i-th and $(i + 1)$-th (a,b) and the corresponding relative trajectories (c,d) for the exchange of the particles i-th and $(i + 1)$-th in the presence of a strong magnetic field for $\nu = 1$ (a,c) and for $\nu = \frac{1}{3}$ (b,d) (third dimension added for better visualization); in both cases, $\nu = 1, \frac{1}{3}$, the cyclotron radius R_c is adjusted to the distance between particles, $2R_c = 2R_0$ and R_0 is sustained by Coulomb repulsion on the level of average separation adjusted to a particle density.

One may add that another idea of real quasi-classical cyclotron collective movement along closed trajectories, in the case of 2D systems with a magnetic field, has been analyzed by Kivelson et al. [Kivelson *et al.* (1986)]. In their paper, the authors considered the free energy input due to the contribution from the coherent collective cyclic rotation of large multiparticle cycles, which is accounted for in the semiclassical approach to the path integral and reveals its particular increase at fractional $1/p$ LLL fillings, which has certain relationships between the analyzed particle arrangement and the Wigner crystal (which was assumed as being the classical equilibrium state). The distant correlations with real quantum dynamics considered in this way were grasped using functional integrals in a quasi-classical approximation. Such collective trajectories have no connection with the considered here purely classical cyclotron braids for purposes of the statistics determination, because the purely classical braids do not reflect real quantum dynamics. However, the cyclic long-range Kivelson correlations clearly point at the closeness of the Laughlin correlations with the particle ordering of the Wigner crystal [Kivelson *et al.* (1986); Lee *et al.* (1987)],

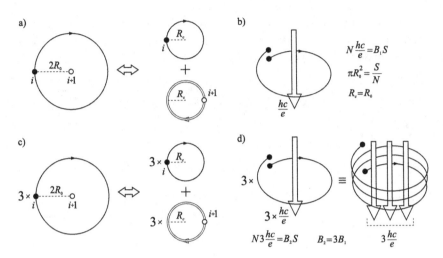

Fig. 4.4   *Double* (and hence closed) relative trajectory of the exchange between the i-th and $(i+1)$-th particles, with one additional loop (i.e., $p = 3$) (a); individual cyclotron trajectories of the exchanging particles the i-th and $(i+1)$-th ones (b); individual cyclotron trajectories corresponding to a double relative trajectory of the exchange of i-th and $(i+1)$-th particles in the presence of a strong magnetic field, for $\nu = 1$ (c) and for $\nu = \frac{1}{3}$ (d), respectively; the number of flux quanta per single particle is marked in both cases $\nu = 1, \frac{1}{3}$ (arrows in (c) and (d)); the resultant effective cyclotron radius R_c fits to the distance between particles, $2R_0 = 2R_c$, allowing particle exchanges in both cases (third dimension added for better visualization).

which is also considered to be a competitive state in terms of composite fermions [Heinonen (1998)].

We again highlight the fact that cyclotron trajectories are not trajectories of real movement; they are merely an auxiliary object for braid group methods, where dynamics play no significant role but topological factors are taken into account. Quantum particles have no trajectories; whereas the braid trajectories reflect the topological, not the dynamical, properties of the system and are related to particle statistics by means of unitary representations.

4.4   Cyclotron structure of composite fermions

The multi-loop character of the relative trajectories of particle exchanges is consistent with the generators of the cyclotron subgroup (4.2) (which also corresponds with the multi-loop form of individual cyclotron trajectories of particles) and is an unavoidable property of charged N–particle

2D systems in the case where the distances between particles (resulting from $\frac{N}{S}$ density and blocked by the Coulomb repulsion) are greater than double the cyclotron radius. This case occurs for fractional $\nu = \frac{1}{p}$ LLL filling; therefore, particle exchanges along individual single cyclotron loops are unavailable. However, there are still other possibilities for exchanges included in the full braid group, which can be gathered in the form of cyclotron subgroup that consists of multi-loop braids. These available exchanges enable the determination of particle statistics through the various one-dimensional unitary representations of the cyclotron braid subgroups. In the case of 2D systems, the multi-loop cyclotron trajectories must share the same common flux of the external field because both the magnetic field and the area of the system do not change; therefore, additional loops take a portion of the flux, or, in other words, there is an effective reduction of the magnetic field strength and the expected growth of cyclotron radii, which again fit into the inter-particle distances (in the case of $\nu = \frac{1}{p}$, p odd, and then cyclotron trajectories with p loops). The loops can only be added to the braids one by one; therefore, one additional loop leads to the generator σ_i^3 (in the simplest case), two additional loops give generators σ_i^5 etc. (but there are no cyclotron groups with the generators σ_i^2, or σ_i^4). The discussed effective weakening of the magnetic field that results from deducting parts of the initial flux of the external field by additional loops in 2D has been modeled by Jain through adding auxiliary flux tube quanta opposite to the external field. These tubes do not actually exist, and the field is not weakened by any auxiliary field flux tubes; however, the field is weakened by the cyclotron loops. This explanation of Jain's flux tubes located on particles does not change the vast usefulness of Jains model for composite fermions in the case of $\nu = \frac{1}{p}$, where p is odd, especially suitable to expanded numerical calculations [Jain (1989, 2007); Heinonen (1998)].

    The total flux of the external field, B, that passes through the area of the system, S, is equal to BS. For p type composite fermions, by considering the *double* exchange of the i-th and $(i+1)$-th particles (hence, along the closed trajectory with $2\frac{p-1}{2} + 1 = p$ loops), we receive the same number of p loops on the individual cyclotron trajectories of the particles (Fig. 4.4) as the number of flux quanta per single particle, $p\frac{hc}{e}$ (when the filling factor is equal to $\nu = \frac{1}{p}$, p odd). Thus, in this case, each loop takes away a single flux quantum, $\frac{hc}{e}$, according to the above formulated rule, because for p type composite fermions, we have p-looped cyclotron trajectories of particles or p flux quanta per particle, $BS = Np\frac{hc}{e}$. On the other hand, the degeneration of each LL is $N_0 = \frac{SBe}{hc}$, (neglecting the spin) and for the fractional LLL filling ν, we get $N_0 = \frac{N}{\nu}$. The equation $\frac{BS}{N} = \frac{hc}{e}\frac{1}{\nu}$ gives $\frac{1}{\nu}$ flux quanta per particle, which is in accordance with the previous estimation for $\nu = \frac{1}{p}$.

In the case of individual cyclotron trajectories with p loops, each loop has its size adjusted to the external field, whose total flux per particle is reduced by the other loops, i.e., by $p - 1$ quanta per particle, precisely the same way as in Jain's model, but only in the case of $\nu = \frac{1}{p}$ where p is odd. Indeed, if $BS = \frac{hc}{e}pN$, then $\frac{hc}{e} = \frac{B}{p}\frac{S}{N}$ and $\frac{B}{p}$ corresponds to a p-times effective field reduction. Based on analogies with the Jain's model, one can argue that for $\nu = \frac{1}{2}$ and $p = 3$, two loops per particle take away the entire flux of the B field, and then the third loop has an infinite radius (Hall metal [Halperin *et al.* (1993)]) in a residual zero field.

Taking away the flux of an external field by adding loops explains the screening of the external field by the model field, which was accepted in Jain's construction with attached auxiliary field flux tubes. The conclusions from Jain's model remain valid also within the approach of cyclotron braids, particularly at $\nu = \frac{1}{p}$ fillings, p–odd (for other fillings, Jain's model turns out to be not so clearly supported in terms of cyclotron braids, as will be presented below).

Let us summarize why the charged particles in 2D systems must be composite particles (this name is slightly confusing and comes from Jain's description, although these composite particles are in fact real quantum particles that have separate statistics and are not composed of any flux tubes) in the case of fields corresponding to fractional LLL fillings. For $\nu = 1$, we have exactly $R_c = R_0$ (where R_c is the cyclotron radius, $\pi R_c^2 B = \frac{hc}{e}$ and $2R_0$ is the distance between particles, that is adjusted to the density and sustained by the short-range part of the Coulomb repulsion, $\pi R_0^2 = \frac{S}{N}$). For $\nu < 1$, the cyclotron radius of normal trajectories without additional loops is $R_c < R_0$; therefore, R_c is *too short* to carry out particle exchanges along these trajectories. However, additional loops lead to the increase in R_c, which again allows exchanges, because, for p-looped cyclotron trajectories, $\frac{hc}{e} = \pi R_c^2 \frac{B}{p}$ and R_c increases compared to the individual (single-looped) cyclotron trajectories; for $\nu = \frac{1}{p}$, again $R_c = R_0$, although the external field is p times stronger than when $\nu = 1$ (at constant N).

Jain's auxiliary flux tubes played a similar role to loops, they increased the cyclotron radius by reducing the effective field. It seems appropriate to conclude that for $\nu = 1$, the cyclotron trajectories are single-looped and that the braid group is a full braid group that is generated by $b_i^{(p=1)} = \sigma_i$, while for $\nu = \frac{1}{p}$, $p > 1$ (p odd), the cyclotron trajectories must be multi-looped, which leads to braids generated by $b_i^{(p)} = \sigma_i^p$.

The main result arising from the braid explanation of the Laughlin correlations is based on proving that composite fermions are not ordinary fermions that are dressed with interaction, in an analogy to quasiparticles in solids, but are statistically separated, real quantum particles. Although wave functions of fermions and composite fermions both are anti-symmetrical, we need to remember about separateness of each of their types and prevent their mixing, just as one does not mix bosons and fermions. This may be important in numerical interaction diagonalizations, which are widely used in the analyses of Hall systems. If all anti-symmetrical trial functions have been included in the Hamiltionian minimization, then the result may be different when compared to being limited to a specific type of composite fermions (with a phase shift of $p\pi$ and not simply π), for which trial functions comprise a subset of an all anti-symmetrical function family; this may be the cause of inaccurate results.

It is similar with numerical analyses of FQHE excitations, which correspond to quasiparticles or quasiholes and are often interpreted as anyons carrying fractional statistics. Phase shifts allow the identification of the statistics, and can be numerically calculated as a Berry phase along a closed trajectory in the configuration space for a model multiparticle wave function that corresponds to low-energy excitations beyond the ground state with fractional Landau level filling. These excitations of quasiparticles or quasiholes are traditionally associated with anyons in the case of a fractional Berry phase. However, it is clear that it is impossible to distinguish between the fractional θ (for the 1DUR of the full braid group) and the fractional $p\alpha$ (for the 1DUR of the cyclotron subgroup) because both of these phase shifts can be the very same fraction. However, because the considered quasiparticles or quasiholes are excitations under a strong magnetic field, the correct states should rather be referred to cyclotron subgroups and composite anyons but not to ordinary anyons, as was commonly accepted previously.

The exchange of anyons for composite anyons may also have consequences for the recently developed topological methods of quantum information processing [Nayak *et al.* (2008); Kitaev (2003); Nielsen and Chuang (2000)]. If we were to approximate any multi-qubit unitary evolution (particularly the universal two-qubit operation) [Nayak *et al.* (2008); Nielsen and Chuang (2000)] using a multi-dimensional irreducible unitary representation (MDUR) (of the appropriate order) of the braid group, then we could implement multi-qubit operations in a physical system of charged 2D particles in a magnetic field. However, for the mentioned approximation to be possible, the crucial condition is that these MDURs must be dense enough in the relevant spaces of the unitary matrices [Nayak *et al.* (2008)]. MDURs may link with degenerated low-energy excitations (quasiparticles or quasiholes) beyond the ground state of the FQHE for fractional LL filling. Because MDUR elements, as matrices, do not commute, the degenerated states are referred to non-abelian anyons (in analogy to abelian anyons, which are associated with the 1DURs of braid groups) [Nayak *et al.* (2008)]. Recent studies of non-abelian anyons, particularly for the low-excited states beyond the 5/2 or 12/5 LL fillings, have shown an insufficient density of MDURs (for non-abelian anyons, in the case of 5/2 filling, the MDURs are insufficient to approximate the required multi-qubit gates [Nayak *et al.* (2008)], whereas the 12/5 state is still unexplained in this respect [Xia *et al.* (2004); Rezayi and Read (2009)]). Therefore, we consider that seeking other favorable fractional statistics systems with more MDURs, which are linked with non-abelian anyons, is of much significance. Taking into account that the subgroups usually have richer representations than the groups, one can expect that the cyclotron braid subgroups would be more useful than the full groups (on account of the higher probability of having beneficial, sufficiently dense representations of the cyclotron braid subgroups).

4.5 The role of the Coulomb interaction

The key role of the short-range part of the Coulomb interaction between particles for generating Laughlin correlations in 2D systems is visible in the fact that the Laughlin function is an *exact* wave function of the ground state for the filling $\frac{1}{p}$ of LLL if we limit the interaction to the first $p-2$ Haldane pseudopotentials [Prange and Girvin (1990); Haldane (1983); Simon *et al.* (2007)]. These pseudopotentials are defined as $V = \sum_{i>j} \sum_m^\infty V_m P_m^{ij}$, where P_m^{ij} is the projector onto the states with the relative angular momentum m of i-th and j-th particles. The Laughlin function turns out to be the exact ground state when V_m is limited to components only with $m = 1, ..., p-2$. These segments of V_m, which express the energy of Coulomb interaction for pairs of particles that have a relative angular momentum $m \leq p - 2$, include the short-range part of the electron interaction, and the remaining segments (i.e., the long-range interaction tail for the larger distances between particles, or for $m = p, ...$) do not substantially affect the Laughlin function [Prange and Girvin (1990); Haldane (1983); Morinari (2000)]. The Laughlin correlations are related to incompressible states, which is reflected by the discrete spectrum of the Coulomb interaction in projection onto Landau states and is expressed in terms of the Haldane pseudopotentials, and is even referred to by Laughlin as the 'quantization of particle separation' [Laughlin (1983a); Prange and Girvin (1990)]. The quantization of the Coulomb interaction following the projection onto the states expressed in terms of the relative angular momentum of pairs in Hillbert subspace corresponding to LLL leads to incompressible FQHE states that are enumerated by integers (eigenvalues of the relative angular momentum of pairs), the same integers that they are in the Laughlin functions (through the exponents in the Jastrow polynomials). It is important to emphasize that, as part of the approach to the FQHE with the use of the Haldane pseudopotentials (i.e., only considering the short-range part of the Coulomb repulsion), the Laughlin correlations can be expressed through the multiparticle Laughlin wave function that corresponds to a precise ground state at fractional LLL filling $\frac{1}{p}$, not a variational result for the state's model [Prange and Girvin (1990); Haldane (1983)]. This highlights the importance of Laughlin correlations (which are expressed by the appropriate phase shift due to particle exchange), which reveals the basic topological property of interacting systems of 2D charged particles under a strong magnetic field. We associate this property with the structure of the cyclotron braid subgroups.

Because of a direct relationship between the Laughlin correlations and the concept of composite fermions, we can state that the Coulomb repulsion (its short-range part expressed in terms of the few initial Haldane pseudopotentials) also has fundamental significance for the construction of composite fermions within its traditional Jain's approach. However, it is clear here that the discrete spectrum of the Coulomb interaction that has energy gaps between states with different relative angular momenta for pairs in the presence of a strong magnetic field (as in the definition of Haldane pseudopotentials) does not allow the treatment of the interaction as a dressing of particles, in analogy to the Landau quasiparticles because of the discontinuity of interaction in this projection.

An effective description method using local gauge field is involved in the Chern–Simons field theory (chiral field, i.e., which breaks the time reversal and parity symmetries). This theory, which was developed earlier for 3D systems, has been introduced into the FQHE area [Fetter *et al.* (1989); Lopez and Fradkin (1991)], and it successfully describes particles with flux tubes modeling any statistics, particularly anyons and Jain's composite fermions. It needs to be added, however, that the Chern–Simons field is accepted *a priori* without providing a physical basis of its specific form, which means that it does not explain the reason for the flux tubes that are attached to particles in the case of composite fermions or anyons (or, equivalently, singular Chern–Simons gauge transformation).

It has been demonstrated that the short-range part of the Coulomb interaction stabilizes composite fermions in the presence of a Chern–Simons field (of its anti-Hermitian part [Rajaraman and Sondhi (1996); Morinari (2000)]), which mixes the states with various momenta inside the LLL [Morinari (2000)] and leads to disturbances in the model of composite fermions within the Chern–Simons field formulation [Heinonen (1998); Morinari (2000)]. The Coulomb interaction removes the degeneration of states with various momenta inside the LLL and causes energy gaps between them that prevent state mixing by the Chern–Simons field, thus stabilizing the composite fermion picture.

Let us also note that for higher Landau levels, the concept of composite fermions may not be as effective as for the LLL because of the possible overlap between the Landau levels that are induced by the interaction [Sbeouelji and Meskini (2001)]. However, in many cases (e.g., for $\nu = 5/2$), the concept of composite fermions turns out very useful [Jain (2007)], especially when the energy gaps between consecutive LLs are larger than the Coulomb interaction, which reduces the mixing of LLs [Heinonen (1998); Jain (2007)]. Analyses (also numerical) show that composite fermions may be used to describe FQHE in the second Landau level [Read (1989); Moore and Read (1991)] including also their paired states [Read and Green (2000); Möller and Simon (2008)]. It was demonstrated that the mixing of LLs contributes to effective interaction of two-body systems, which stabilizes the so-called Moore-Read state which is expressed via the Pfaffian factor [Bishara and Nayak (2009); Rezayi and Simon (2011)] (cf. Chapter 7.2).

The short-range part of the Coulomb interaction also stabilizes the picture of composite fermions in the framework of the presented braid approach, similarly as it removes the momentum orbit instability inside the Landau level caused by the Chern–Simons field [Morinari (2000)]. Indeed, if we reduce the short-range part of the Coulomb repulsion, the distances between the particles might not be rigidly kept (and would only be determined in average according to the density) and then other cyclotron trajectories, in addition to the multi-loop trajectories (for $\nu = \frac{1}{p}$), would be allowed, which would cause violation of the cyclotron structure of the braid subgroup.

Therefore, the short-range part of the Coulomb interaction is crucial in every approach for composite fermions. Restricting the full braid group to its cyclotron subgroup, with multi-loop trajectories, is justified only when the inter-particle separation is adjusted to the cyclotron radius (of multi-looped cyclotron orbit). The role of the Coulomb repulsion at close distances is to protect against the excessive approach of the particles and to maintain the almost rigid distance between them at a level that results from the density. Thus, the short-range part of the Coulomb interaction is employed to construct cyclotron subgroups. The long-range part of the Coulomb interaction is left out as the particle residual interaction, what is similar to the model of Jain's composite fermions, which are almost free particles with a weak and long-range residual interaction [Jain (1989); Heinonen (1998); Jain (2007)].

4.6 Composite fermions in terms of cyclotron groups

Jain's model of composite fermions assumes [Jain (1989); Heinonen (1998); Jain (2007)], that every particle has a $p - 1$ flux tube quanta attached to it, for p type composite fermions. It is assumed that these local flux tubes are oriented opposite to the external field. Then, the weakening of the external field by the mean field of these local tubes takes place, which can by estimated by calculating the reduction of the external flux caused by the flux tube,

$$BS - N(p-1)\frac{hc}{e} = \pm B'S, \qquad (4.4)$$

where $\pm$ stands for the assumed possible orientation of the reduced resultant B' field, along or opposite to the external field B orientation. In a reduced resultant field B' IQHE can take place, i.e.,

$$B'S/\frac{hc}{e} = \frac{N}{n},$$

where n indicates the LL number (the number of completely filled Landau levels, all of which have the same degeneration, and which depends on the strength of the field, equal to $B'S/\frac{hc}{e}$). The initial (real) external field corresponds to the fractional LLL filling,

$$\nu = \frac{N}{N_0} = N/BS/\frac{hc}{e}.$$

The above relations give us, $BS = N/\nu\frac{hc}{e}$ and then $N/\nu\frac{hc}{e} - N(p-1)\frac{hc}{e} = \pm N/n\frac{hc}{e}$, or $1/\nu - (p-1) = \pm 1/n$, therefore:

$$\nu = \frac{n}{n(p-1) \pm 1},\qquad(4.5)$$

which results in the main sequence of ν fillings for the FQHE in a real external field (expressed by the incompressible states of full filling [separated by gaps, if the Coulomb interaction does not reduce them] of subsequent n Landau levels in a reduced resultant field), owing to the $p-1$ flux tube quanta that are attached to the particles in the case of p-type composite fermions.

In the case of two flux tube quanta that are attached to Jain's composite fermions, i.e., for $p = 3$, we will obtain the field $B' = 0$ for $\nu = \frac{1}{2}$. Indeed, in such a case,

$$B'S = BS - \frac{1}{2}N_0 2\frac{hc}{e} = BS - \frac{BS}{\frac{hc}{e}}\frac{hc}{e} = 0.$$

This is interpreted as a state of the Hall metal [Halperin *et al.* (1993)], where composite fermions and their local flux tubes reduce the whole flux of the external field, and the resulting composite particles behave as fermions without the presence of any field, they create the Fermi sea, as in metal [Halperin *et al.* (1993); Jain (2007)].

However, the phenomenological scheme of composite fermions presented above is not fully convincing if we accept the explanation that the attached flux tube quanta do not exist, but are merely model substitutions of additional cyclotron loops. We should particularly verify the possibility described above, when a reduced resultant B' field may have an opposite direction to the external field B (i.e., the minus sign in the accepted $\pm$ in the equation (4.4)). Moreover, it seems reasonable that the loops cannot reduce the external flux beyond its full value. Therefore, it seems that the possibility of local flux tubes 'prevailing' over the total external flux, as is assumed in Jain's model, is not clear. We will demonstrate below that by improving this assumption in terms of the cyclotron braid groups, we can also obtain a hierarchy of the FQHE fillings through a relation with IQHE and justify the state of Hall metal. To do this, we need to note the following:

- only for $\nu = \frac{1}{p}$, where p odd natural number, each cyclotron loop can be assigned with a flux quantum; e.g., for $p = 3$ there are two additional flux quanta per particle (i.e., three flux quanta per particle in total); let us add that cyclotron loops are not real particle trajectories and we can only use the coincidence of the number of BS flux quanta per loop with the same number of flux quanta per particle, and in the case of p loops and the field corresponding to the filling $\nu = \frac{1}{p}$ we deal with a full quantum $\frac{hc}{e}$ per loop;

- for other fillings ν of LLL, the flux per loop is $\phi = \frac{BS}{pN} = \frac{BS}{p\nu N_0} = \frac{BS}{p\nu\frac{BS}{hc/e}} = \frac{hc}{e}\frac{1}{p\nu}$; it is some fraction of the full quantum; if $\phi = \frac{hc}{e}\frac{1}{n}$, as in the case of n fully filled Landau levels (with single cyclotron loops per particle), we may expect a similar relationship between the FQHE and IQHE (incompressible state), as in Jain's model, and thus we get the hierarchy of FQHE fillings, $\frac{1}{n} = \frac{1}{p\nu}$, or,

$$\nu = \frac{n}{p}, \quad p = 3, 5, 7, \ldots \ n = 1, 2, 3, 4, \ldots \tag{4.6}$$

In this way, we can reproduce all fractions given in the relation (4.5) without attaching full flux tube quanta outside the filling $\nu = \frac{1}{p}$ (p odd) and without assuming the possibility of changing the direction of the reduced resultant field;

- in the case of $\nu = \frac{1}{2}$ and $p = 3$ we obtain $\phi = \frac{BS}{3N} = \frac{eh}{c}\frac{2}{3}$; therefore, tree loops take two full flux quanta and we may phenomenologically presume that the dynamics of the quantum system will reflect the triple cyclotron trajectories, two of loops will take away the entire flux of the external field and the third one, without the field, realizes Fermi's liquid; we would once more point out that the cyclotron loops (associated with the classical braid image) do not have to embrace the exact quantum of flux as opposed to *quasi-classical* trajectories, to which the quantization of the magnetic field flux rigidly refers; with regard to Hall systems with arbitrary filling ratio of LL we can only presume quasi-classical character of related wave-packet electron orbits, but only in the cases when it is possible, as for e.g., $\nu = \frac{1}{2}$ (i.e., when classical cyclotron trajectories are assigned with integer multiples of the flux quantum). It is also worth mentioning that for the full filling of higher Landau levels, each particle is assigned with a $\phi = \frac{hc}{e}\frac{1}{n}$ fragment of flux quantum and that it is impossible for an individual particle to undergo individual quasi-classical cyclotron movement (which would need the entire quantum of flux piercing the quasi-classical orbit); rather, its collective arrangement that does not oppose the quantization of the flux is rigidly required in the quasi-classical quantum description.

With regard to classical cyclotron loops (without the need of flux quantization) that are associated with classical braid trajectories, the number of loops is always an integer, whereas the number of flux quanta per each loop is not always an integer (as in the filling apart $\nu = \frac{1}{p}$).

4.7   Hall metal in the description of cyclotron groups

Let us refer to the state for $\nu = \frac{1}{2}$ (called the Hall metal) from the viewpoint of braid groups. In Jain's model of composite fermions, two flux tube quanta are attached to particles to completely erase the external field in approximation of the mean field (in other words, the Hermitian field of Chern–Simons, which is associated with Jain's model cancels the whole external field within the mean field approximation; this leads to the state of composite fermions without the field, i.e., the Fermi sea, here referred to as the Hall metal [Halperin *et al.* (1993)]. In Read's approach to composite fermions when $\nu = \frac{1}{2}$, a complete screening of charge takes place because of the depletion of the charge density near the middle of the vortex and because of the screening by a positive unbalanced background charge (*jellium*), for the vorticity $q = 2$ and filling fraction $\nu = \frac{1}{2}$. The exchange of two vortices generates a phase shift of $q^2\nu\pi = 2\pi$ that is supplemented with π when we exchange electrons together with the vortices attached. Therefore, complexes composed of two vortices that are attached to electrons behave like fermions (with no charge), and they create the Fermi sea (Hall metal). The instability of a normal system of fermions then leads to a paired BCS state [Abrikosov *et al.* (1975)], which is expressed in the position representation by the Pfaffian term (cf. Chapter 5.4). The state with the pairing reproduces the incompressibility because of the pairing energy gap (e.g., considered for $\nu = 5/2$ [Moore and Read (1991); Willett *et al.* (1987)], also taken into account for $\nu = 1/2$ and $1/4$ [Luhman *et al.* (2008); Shabani *et al.* (2009a)]). Since Pfaffian [Greiter *et al.* (1992)] produces a phase shift of $-\pi$ when particles exchange, then the full phase shift of the wave function with the Jastrow factor $\prod_{i>j}(z_i - z_j)^2$ [Greiter *et al.* (1992); Moore and Read (1991)] is π. Such a phase shift is defined by the 1DUR of the cyclotron group (with $p = 3$ because this subgroup corresponds to the region of fillings $\nu \in [1/3, 1)$) that are marked by $p\alpha = 3\frac{1}{3}\pi = \pi$, or, $\alpha = \frac{1}{3}\pi$. This representation ($p = 3, \alpha = \frac{1}{3}\pi$) induces a fermion statistic for $\nu = 1/2$ [Jacak and Jacak (2010)] and, in terms of braid composite fermions, corresponds to the Fermi sea (because two loops take away the

entire external field flux) in accordance, on the other hand, with the local gauge transformation with $q = 2$, reproducing fermions (starting with
ordinary fermions) [Read (1994); Rajaraman and Sondhi (1996)].

4.8   Comments on restrictions for the multi-loop structure of cyclotron braids

Quantization of the flux of a magnetic field is an extremely important result;
obligatory regardless of the interaction and other system particularities, it
has a quasi-classical character, where the quantum dynamics generally not
going along a trajectory can be characterized (with precision admitted by
quasi-classical approximation) by the location and momentum of an appropriate wave packet to allow the determination of the orbit and its area
and, hence, the field flux passing this area [Abrikosov *et al.* (1975)]. The
cyclotron trajectories that are considered above, which are associated with
braids, are not real particle trajectories, even in the quasi-classical sense.
Quantum dynamics does not have the character of movement along trajectories; moreover, in the case of systems in a magnetic field, the shape
of the wave function is strongly dependent on the choice of a gauge of the
field vector potential [Landau and Lifshitz (1972); Eliutin and Krivchenkov
(1976)]. However, the periodicity of the quasi-classical image [Eliutin and
Krivchenkov (1976)] in the presence of a magnetic field (any quasi-classical
wave packet undergoes periodic dynamics on a plane perpendicular to the
magnetic field, as presented below) may suggest the usefulness of the quasi-
classical description in relation to the quantization of the magnetic field
flux, which is valid in reference to quasi-classical orbits. In this case, when
part of the external field flux per a single cyclotron loop is expressed by
the full flux quanta, it seems possible that the quasi-classical image of a
strongly interacting system of charges in a magnetic field can reflect the
cyclotron structure as well. Such a completely heuristic presumption may
combine the topological braid structure with real dynamics in these specific
situations that are pointed out by the quasi-classical flux quantization. This
may especially enable us to determine the restrictions that are imposed on
the usefulness of the cyclotron multi-loop structure by the dynamic requirements of minimizing the kinetic energy and interactions. Such restriction
is also found in Jain's standard theory of composite fermions, which shows
[Jain (2007); Heinonen (1998)], that for $\nu < \frac{1}{9}$, the state of the Wigner
crystal is more stable than the Laughlin state. For a magnetic field of a

magnitude that corresponds to so low filling fraction, the cyclotron radius is very small and a braid description would require a structure with as many as 9 loops, which, in the face of the suggested possibility of dynamic consequences, can turn out to be energetically inconvenient. In such case, we will be left with a simple, single-loop cyclotron movement, which would not allow particle exchanges and establishing statistics. Electrons will be located within a structure of the Wigner's triangular planar network. Investigation of the quasi-classical trajectories, which are performed in papers by [Kivelson *et al.* (1986); Lee *et al.* (1987)] clearly confirm this concept of Wigner crystallization.

The Wigner crystal, or the electron crystal [Wigner (1934)], corresponds to the crystallization of the electron liquid, where the size of the packet that corresponds to the particle localization is smaller than the average distance between particles. Such crystallization can be expected at low densities (and at relatively large average distances between particles), when the competition between the increasing kinetic energy, as a result of localization (opposes localization), and the interaction (prefers localization) may enable localization (in higher densities kinetic energy dominates). The state of the Wigner crystal was observed in a two-dimensional system of electrons on the surface of liquid helium [Grimes and Adams (1979)]. The magnetic field enhances localization, and in 2D Hall electron systems at low fillings, we may also expect Wigner crystallization to compete against Laughlin collective states. By comparing the total energy value (kinetic and interaction energy) of the state described by Laughlin functions with the state of the electrons localized on a Wigner crystal network in planar 2D systems, the critical LLL filling $\nu = \frac{1}{10}$ has been determined; below this value, the Wigner crystal turns out to be more energetically favorable [Levesque *et al.* (1984)]. Taking into account the mixing of states between the Landau levels, this moves the border filling rate to $\nu = \frac{1}{9}$ [Mandal *et al.* (2003)]. A numerical diagonalization of a system of 6 particles on a torus [Yang *et al.* (2001)] points to a critical filling of $\nu = \frac{1}{8}$. The reason for these slight discrepancies can be sought in the close energy position of both competing states in a relatively wide area of various parameters, which have been accounted for to various degrees using different calculation approaches. Experimental confirmation of the realization of the Wigner crystal in a 2D electron system in a strong field has not been achieved thus far; nevertheless, FQHE observations at low LLL fillings have not been excluded as well [Jain (2007)].

4.8.1 *Periodic character of wave packets' dynamics*

The classical cyclotron trajectory of a charged particle has a circular shape on a 2D plane (more generally speaking, on a plane perpendicular to a field in 3D). The quantum equivalent of these dynamics may (depending on the choice of a field gauge) not even display axial symmetry (e.g., in the Landau gauge such symmetry is not exhibited by Landau states with Hermite functions with respect to one of the coordinates on a plane perpendicular to the field [Landau and Lifshitz (1972)]). This results from the degeneration of the Landau levels, which in a classical case, is expressed by lack of a determined symmetry axis of the trajectory.

For the gauge [Eliutin and Krivchenkov (1976)], $\mathbf{A} = (-\frac{1}{2}By, \frac{1}{2}Bx, 0)$, the classical Hamilton function for a single particle has the following form:

$$H = \frac{1}{2m}\left(p_x - \frac{|e|}{2c}By\right)^2 + \frac{1}{2m}\left(p_y + \frac{|e|}{2c}Bx\right)^2 + \frac{p_z^2}{2m}$$

and the Hamilton equations (the convention $|e| = e$ was adopted here),

$$\dot{x} = \frac{1}{m}\left(p_x - \frac{eB}{2c}y\right), \; \dot{y} = \frac{1}{m}\left(p_y + \frac{eB}{2c}x\right), \; \dot{z} = \frac{1}{m}p_z,$$
$$\dot{p}_x = -\frac{\omega}{2}\left(p_y + \frac{eB}{2c}x\right), \; \dot{p}_y = \frac{\omega}{2}\left(p_x - \frac{eB}{2c}y\right), \; \dot{p}_z = 0.$$

The solution to these equations is as follows: $x = Rcos(\omega t + \psi) + x_0$, $y = Rsin(\omega t + \psi) + y_0$, (where $\omega = \frac{eB}{mc}$ is the cyclotron frequency, R is the cyclotron radius, and x_0, y_0 is the position of the orbit's center). Therefore, $y_0 = \frac{y}{2} + \frac{p_x}{m\omega}$, $x_0 = \frac{x}{2} - \frac{p_y}{m\omega}$ and replacing the canonically coupled variables x, p_x, y, p_y with operators, we obtain the commutator, $[\hat{x}_0, \hat{y}_0] = \frac{i\hbar}{m\omega}$, which means that both coordinates of the orbit's center cannot be simultaneously determined. Analogically to the classical solutions, we can define the operators $\hat{x}$, $\hat{y}$ in the Heisenberg representation [Eliutin and Krivchenkov (1976)]:

$$\hat{x}(t) = \hat{A}cos\omega t - \hat{B}sin\omega t + \hat{x}_0,$$
$$\hat{y}(t) = \hat{A}sin\omega t + \hat{B}cos\omega t + \hat{y}_0,$$

where operators independent of time, $\hat{A} = \frac{\hat{p}_y}{m\omega} + \frac{\hat{x}}{2}$, $\hat{B} = -\frac{\hat{p}_x}{m\omega} + \frac{\hat{y}}{2}$ and $\hat{x}_0 = \frac{\hat{x}}{2} - \frac{\hat{p}_y}{m\omega}$, $\hat{y}_0 = \frac{\hat{y}}{2} + \frac{\hat{p}_x}{m\omega}$ (with the initial condition, $\hat{x}(0) = \hat{x}$, $\hat{y}(0) = \hat{y}$). These relations show that the position operators in the Heisenberg representation depend on time in a periodic way, and it therefore follows [Eliutin and Krivchenkov (1976)] that any wave packet of any shape on a plane x, y will also vary in time periodically with the period of $2\pi/\omega$. This provides some idea about the quantum cyclotron behavior of a particle in a magnetic field, particularly in a quasi-classical approximation, where a position-localized wave packet can be identified with a moving quasi-classical particle. Based on the above, we can expect a relationship between the real quantum dynamics and a cyclotron braid image.

We may note the fact that the wave packet that corresponds to the ballistic, quasi-classical dynamics of a particle is closely related to the collective character of a multiparticle system, especially when the momentum enumerates the single-particle stationary states, which may then produce ballistic wave packets. The collective movement minimizes the kinetic energy, while the interaction favors localization (and causes the related increase in kinetic energy). Therefore, the collective dynamics seems to prefer the quasi-classical movement of ballistic packets along periodic closed trajectories, which necessarily embrace quantized external magnetic field fluxes. This suggests the role of collectivization in the energy preference of wave packets realizing closed trajectories in reference to the classical cyclotron description, including the multi-looped braid picture. It seems to be in accordance with the FQHE observations in graphene (described in the following paragraph), which is found at low carrier density, and, hence, accompanies their dilution and the resulting reduction of interaction. Perhaps the interaction does not play such a major role in initiating the FQHE, as previously expected in view of the standard model of composite fermions, which treated the dressing of fermions with localized flux tubes as a result of just the interaction itself [Heinonen (1998); Jain (2007)].

Experimental confirmation of the increased size of cyclotron orbits of 2D electrons in a strong magnetic field can be found in the readouts of the resonance of the layer's piezoelectric coupling with surface acoustic waves, and the system of 2D electrons placed below it. Resonance energy transfer from the acoustic wave to the electron system was observed when the acoustic wavelength is commensurate with the cyclotron radii corresponding to composite fermions in a strong perpendicular magnetic field [Willett *et al.* (1993, 1995)]. Another experimental result suggesting that the increased size of the cyclotron trajectories at fractional LLL fillings is observed in the ballistic transport of carriers in a 2D system in a strong field between two nano-slots placed horizontally. Cyclotronic focusing of transport was observed where the distance between holes (slots) was commensurate with the cyclotron radius length, suggesting the cyclotron concentration of carriers along quasi-classical packets orbits [Fleischmann (1996)].

4.8.2 Quasi-classical character of quantization of the magnetic field flux

Generalized momentum [3] in the presence of a magnetic field is expressed by the equation [Olchovskii (1974)],

$$\mathbf{P} = \mathbf{p} + \frac{e}{c}\mathbf{A}, \qquad (4.7)$$

where $\mathbf{p}$ is the kinematic momentum. We change the momentum operator $\hat{\mathbf{p}} = -i\hbar\nabla$ to the operator $\mathbf{P} - \frac{e}{c}\mathbf{A}$, which leads to the single-particle Hamiltonian $H = \frac{(-i\hbar\nabla - \frac{e}{c}\mathbf{A})^2}{2m} + U$ (we exchanged generalized momentum $\mathbf{P}$ with the operator $-i\hbar\nabla$, as a canonical variable according to the appropriate Poisson brackets, which is in compliance with the principles of commutation).

Choosing the Landau gauge, $\mathbf{A} = (0, Bx, 0)$, we have for the kinematic momentum, $p_x = -i\hbar\frac{\partial}{\partial x}$, $p_y = -i\hbar\frac{\partial}{\partial y} - \frac{e}{c}Bx$; hence,

$$p_y p_x - p_x p_y = -i\hbar\frac{e}{c}B, \qquad (4.8)$$

or

$$p_y Y - Y p_y = -i\hbar, \qquad (4.9)$$

where, $Y = \frac{c}{eB}p_x$. Therefore, (p_y, Y) can be regarded as a pair of canonically coupled variables, and with the use of Bohr-Sommerfeld formula for quasiclassical approximation,

$$\oint p_y dY = h(n + \gamma(n))q, \qquad (4.10)$$

where γ is a slow varying function of n, $0 < \gamma(n) < 1$, whereas q is the number of loops of the classical closed (cyclotron) trajectory in the case of 2D [4]. Therefore,

$$\oint p_y dp_x = \frac{heB}{c}(n + \gamma(n))q, \qquad (4.11)$$

---

[3]In classical mechanics, generalized momentum $\mathbf{P} = \frac{\partial \mathcal{L}}{\partial \mathbf{v}}$, and for a particle in a magnetic field, Lagrangian has the following form: $\mathcal{L} = \frac{mv^2}{2} + \frac{e}{c}\mathbf{A} \cdot \mathbf{v}$; Hamilton equations are expressed for the generalized momentum (generalized momentum and generalized position make a pair of canonical variables, appropriate for quantization), the Hamilton function has the form $H = (\sum P\dot{q} - \mathcal{L})_{\dot{q}(q,P,t)} = \frac{(\mathbf{P} - \frac{e}{c}\mathbf{A})^2}{2m} + U$; however, the kinematic momentum, $\mathbf{p} = m\mathbf{v}$, is associated with a force, $\mathbf{F} = \frac{d\mathbf{p}}{dt} = m\frac{d\mathbf{v}}{dt}$, in particular, the Lorentz force, $\mathbf{F} = \frac{e}{c}q\mathbf{v} \times \mathbf{B}$.

[4]In the case of 2D and multi-looped classical cyclotron trajectories in a configuration space; in accordance with the braid subgroup, we add the q multiplier based on the quasiclassical relation [Abrikosov et al. (1975)], $\mathbf{F}dt = d\mathbf{p} = \frac{e}{c}d\mathbf{r} \times \mathbf{B}$ ($\mathbf{F}$ Lorentz force) the position space trajectory is replicated, within a quasi-classical approximation, in the kinematic momentum space (orthogonally rotated and scaled with the $\frac{c}{eB}$ factor), which means it should also be multi-looped.

or 'area quantum' in the space of kinematic momentums is $\Delta S_p = \frac{heBq}{c}$.

The area of the quasi-classical trajectory in a space of kinematic momentum is therefore scaled in the same way as Bq and similarly to the kinetic energy (being proportional to square kinematic momentum). Thus, we can expect that, along with the increasing q, the quantum equivalent of a multi-looped cyclotron structure is also becoming less and less energetically convenient, which is due to the rapid increase in kinetic energy (potential energy is constant according to the adopted condition of constant density and homogeneity of the system). A state with $q = 1$ at fractional fillings (i.e., in high fields) cannot be a collective multiparticle state (because the single-loop cyclotron trajectories are too short); therefore, Laughlin states are implemented. However, when the $q \geq 9$ energy advantage of the collective Laughlin states disappears, and the state of the individual localized particles (with undefined statistics, as the particles are unable to exchange under such a strong magnetic field) becomes more energetically convenient [5], clearly suggesting creation of the Wigner crystal.

With regard to Jain's assumption on the total number of flux tube quanta that are attached to each particle, even for fillings outside $\nu = \frac{1}{p}$, we may observe from the viewpoint of the multi-looped braid structure of cyclotron trajectories that such a requirement would have to relate to the abovementioned possibility to replicate the multi-loop braid structure by the real cyclotron dynamics of wave packets. According to the above, every wave packet undergoes periodic dynamics (with cyclotron period); in particular, the ballistic packet would have to describe closed trajectories, which must embrace quantized external field fluxes. In the case of $\nu = \frac{1}{p}$, all loops of a multi-looped cyclotron trajectory would include in this image the full flux quanta, but in the case of ν, apart from $\frac{1}{p}$, the last loop could not embrace the full quantum (the rest of the flux after lowering it by previous flux quanta may be either a positive or negative fraction of the flux quantum, depending on the direction of the resultant Jain field with respect to the direction of the external field).

What seems interesting is the observation that maintaining the assumptions of the standard theory of composite fermions, in the case also apart of $\nu = \frac{1}{p}$, based on the model with the integer number of flux quanta attached to each particle, in a proposed multi-looped interpretation leads to the conjecture that part of those loops would embrace full flux quanta, and only

---

[5] For electron wave functions localized on the Wigner triangular network, described by the so-called Maki-Zotos wave-functions [Maki and Zotos (1983)], the overlap integral for the nearest neighbors reaches $\sim e^{-3.6/\nu} \sim e^{-30} \sim 0$, for $\nu = 1/9$.

the last one will embrace a fragment of the quantum; this fragment may be positive or negative, depending on whether Jain's effective resultant field is directed along with or opposing the external field (i.e., in the case of $p = 3$ for $\nu < \frac{1}{2}$ or $\nu > \frac{1}{2}$, respectively). Such situations would fit real closed periodic trajectories of the quasi-classical wave packets, for which quantization of the field flux is absolutely obligatory. This would mean that in the case, where the last loop embraces a n-th part of the flux quantum (as in the case of fully filled n Landau levels), it would be possible to arrange the quasi-classical multi-looped real movement. One can presume that the last loop would be organized analogically to the n Landau level, collectively, i.e., with the collective participation of n particles, which only together are able to take away the full quantum of a magnetic field flux (each $\frac{1}{n}$ fraction of the quantum). Such a collective loop might be represented as a loop, whose circumference contains n de Broglie waves (each of them somewhat represents a single particle). This trajectory would carry out the cyclotron movement according to the direction of previous loops or opposite to it, depending on the direction of the residual resultant Jain's field with respect to the direction of the external field. In the case of flux quantization that forces such a reversed cyclotron movement of the last trajectory, its relation with the previous loops would be shaped like the number eight. Such an exotic trajectory of possible multi-loop quantum dynamics perceived in the categories of periodic movement of ballistic wave function would result from the quasi-classical flux quantization, which perhaps is being carried out. The good compliance between the model of composite fermions with FQHE) seems to confirm that picture, especially near $\nu = \frac{1}{2}$), but only for fillings included in the $\nu = \frac{n}{n(p-1)\pm 1}$ hierarchy. Apart from these fillings, the arrangement of quasi-classical trajectories of wave packets is impossible owing to flux quantization constraints. As previously described, an alternative approach toward the FQHE mapping to IQHE, with identical flux fraction per each of the loops of a multi-looped structure (in a classical braid approach, without having to quantize flux), leads to a slightly modified hierarchy of fillings, $\nu = \frac{n}{p}$. The experimental observation of the Hall metal properties for $\nu = \frac{1}{2}$ seems to suggest the advantage of the previous model.

One may plan an experiment intended to determine which of the above interpretations is correct. The experiment might be based on measuring the cyclotron focusing of a 2D beam of carriers that passes through narrow, nanometer-sized slots, as in experiments described in [Fleischmann (1996);

Heinonen (1998)]. By measuring the focusing to the left and right of the source slot in a nearby control slots, we could detect orientation changes in the cyclotron movement on the last orbit of the multi-loop trajectory, which would be strongly elongated near $\nu = \frac{1}{2}$. A change in the movement orientation along this orbit would lead to the observation of asymmetry while passing through the filling $\frac{1}{2}$, when changing orientation of the resultant Jain's field with respect to the external field.

4.9   Cyclotron groups in the case of graphene

A single-atom thick layer of graphite (an allotropic form of carbon), known as graphene, creates a hexagonal 2D structure with a Bravais lattice that has two vectors:

$$\vec{a}_1 = a(3, \sqrt{3})/2, \;\; \vec{a}_2 = a(3, -\sqrt{3})/2,$$

($a \simeq 0.142$ nm, distance between carbon atoms) with two carbon atoms per unit cell, Fig. 4.5 (a).

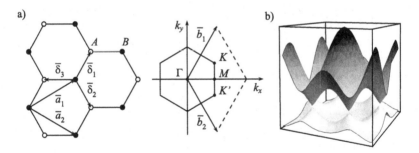

Fig. 4.5   a) A two-sublattices (A, B) triangular structure of graphene, $\vec{a}_1$, $\vec{a}_2$ are Bravais lattice vectors, $\vec{b}_1$, $\vec{b}_2$ are vectors of reciprocal lattice, b) the band structure of graphene, band type π, in a model of strong coupling, in compliance with the relation, $E_{\pm}(\mathbf{k}) = \pm t\sqrt{3 + f(\mathbf{k})} - t' f(\mathbf{k})$, where $f(\mathbf{k}) = 2\cos(\sqrt{3}k_y a) + 4\cos(\frac{\sqrt{3}}{2}k_y a)\cos(\frac{3}{2}k_x a)$, $t = 2.7$ eV hopping energy to the nearest neighbors (between sublattices, vectors $\vec{\delta}_i$), $t' = 0.2t$ hopping energy to next-nearest neighbors (inside the sublattices), $a = 1.42$ Å, the ideal Dirac points K and K' for $t' = 0$.

This results in the creation of a double triangular lattice, which is a hexagonal structure of carbon atoms that are arranged in a honeycomb pattern, an outspread nanotube. Hybridized bonds sp^2 of the σ type lead to a strong (covalently bonded) two-dimensional structure, whereas p orbitals, perpendicular to the plane, hybridize to type π of the band structure (which is well described in the approximation of a strong coupling and with the inclusion of nearest neighbors and subsequent neighbors) with the hole valley (valence band) and the electron valley (conduction band) meeting at points K and K' at the border of a hexagonal Brillouin zone [Wallace (1947); Castro Neto *et al.* (2009)], Fig. 4.5 (b). Both bands meet at these points (non-gap semiconductor) and have a conical shape (in the case of $t' = 0$, near points K, K'), which means that the relationship between the energy and the momentum (distance from points of contact) is linear with respect to momentum length. The appropriate band Hamiltonian (within the strong coupling approximation including the nearest neighbors and accounting for both sublattices, numerated with an artificially introduced pseudospin) is formally equivalent to the description of relativistic fermions with zero rest mass ($E = \pm\sqrt{m_0^2 v_F^4 + p^2 v_F^2}$, with $m_0 = 0$), described by the Dirac equation with the velocity of light replaced by the Fermi velocity, $v_F \simeq c/300$ [Castro Neto *et al.* (2009); Yang (2007)]. Therefore, the dynamics equation is as follows:

$$-iv_F \vec{\sigma} \cdot \nabla \Psi(\mathbf{r}) = E\Psi(\mathbf{r}),$$

where the Pauli matrix vector corresponds to the pseudospin structure that is related to two sublattices [Castro Neto *et al.* (2009); Geim and MacDonald (2007)] (wave functions are spinors in this structure). The zero mass of the Dirac fermions leads to numerous consequences and electron anomalies in the properties of graphene [Castro Neto *et al.* (2009); Geim and MacDonald (2007); Novoselov *et al.* (2005); Zhang *et al.* (2005)]. For Dirac particles with zero rest mass, the momentum uncertainty also leads to energy uncertainty (in contrast to the non-relativistic case, where the relationship between the uncertainty of position and momentum is independent of the relationship between the uncertainty of energy and time), which leads to the time evolution mixing together particle states with hole (anti-particle) states for relativistic type dynamics. For zero-mass Dirac electrons, the scaling of cyclotron energy is different as well ($\sim B^{1/2}$, and not $\sim B$, as in the case of non-relativistic particles). The value of this energy is also different and larger by far (two orders of magnitude larger than the corresponding one in classical materials, i.e., it is [owing to zero mass in the Dirac point] as high as approximately 1000 K, for a 10 T field), which allows us to observe the Integer Quantum Hall Effect in graphene, even at room temperature [Novoselov *et al.* (2005); Zhang *et al.* (2005)]. There is, however, an anomalous IQHE observed here (for $\nu = \pm 4(n + 1/2)$, or for $\pm 2, \pm 6, \pm 10, \ldots$ and at the zero Landau level in the Dirac point, i.e.,

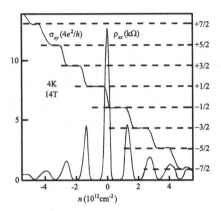

Fig. 4.6   IQHE in graphene as a function of concentration (controlled with lateral gate voltage): the peak for $n = 0$ indicates the existence of Landau level for the Dirac point; electron and hole bands lead to symmetric IQHE oscillation; plateaus σ_{xy} correspond to half multiplicities of $4e^2/h$, according to the structure of double-valley pseudospin. *Source: A. H. Castro Neto, F. Guinea, N. M. R. Peres, K. S. Novoselov, and A. K. Geim, "The electronic properties of graphene," Rev. Mod. Phys. 81(1), p. 109, 2009.*

for zero energy; $\pm$ corresponds to particles and holes, respectively, 4 results from pseudospin/valley degeneration, $1/2$ is associated with Berry's phase for Dirac particles) [Novoselov *et al.* (2005); Zhang *et al.* (2005); Yang (2007)]); cf. Fig. 4.6, which is well-explained by the band structure leading to an effective Dirac description [Castro Neto *et al.* (2009); Geim and MacDonald (2007); Novoselov *et al.* (2005); Zhang *et al.* (2005); MacClure (1956)]. The so-called Klein paradox, which refers to the ideal tunneling of Dirac particles by rectangular potential barriers, leads to extensive mobility of the charge carriers in graphene, which is experimentally observed even near the Dirac point (the Fermi level at the border between electrons and holes). At this point, the density of charges is zero (and the zero Landau level is located here, employing both bands) [Castro Neto *et al.* (2009); Novoselov *et al.* (2005); Zhang *et al.* (2005); Yang (2007)]. Graphene is therefore a material with unique properties, which have only been minimally characterized both experimentally and theoretically as yet. A. Geim and K. Novoselov were awarded the Nobel Prize 2010 for developing a method for obtaining graphene planes and describing the numerous properties of graphene.

Searching for states related to the FQHE in the case of Hall graphene measurements is particularly interesting. Despite the use of very strong magnetic fields (up to 45 T), the FQHE was not detected in graphene samples that were deposited onto a substrate of SiO_2 [Zhang *et al.* (2006)]. In this paper, however, the emergence of additional plateaus of IQHE for the fillings $\nu = 0, \pm 1, \pm 4$ was noted, indicating the elimination of spin-pseudospin degeneration (related to sublattices) as a result of increasing mass of the Dirac fermions [Zhang *et al.* (2006)]. Only after mastering the novel technology of so-called suspended ultrasmall graphene scrapings with extreme purity and high mobility of carriers (above 200000 cm^{2}V^{-1}s^{-1}; high mobility is also necessary to observe FQHE in the case of semiconductor 2D hetero-structures, which may be related to multi-looped quasi-classical cyclotron movement of wave packets in the case of multi-looped braids associated with FQHE) was it possible to observe the FQHE in graphene at fillings $\nu = 1/3$ and $-1/3$ (the latter for holes, with opposite polarization of the gate voltage, which determines the position of the Fermi level, either in the conduction band or in the valence band) [Du *et al.* (2009); Bolotin *et al.* (2009)]. Both papers report the observation of the FQHE in graphene under strong magnetic fields. The paper [Bolotin *et al.* (2009)] reports the FQHE in a field of 14 T, for an electron concentration of 10^{11}/cm^2; the paper [Du *et al.* (2009)] reports the FQHE in a field of 2 T but for a concentration level that is one order of magnitude lower Fig. 4.7.

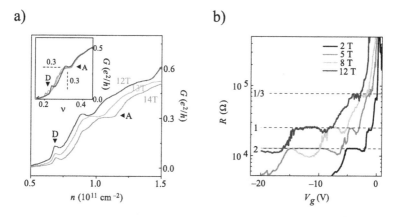

Fig. 4.7   a) FQHE observation in suspended graphene for the filling 0.3 (1/3) in a field of 12-14 T with a concentration of 10^{11}cm^{-2} and a mobility of 250000 cm^{2}V^{-1}s^{-1}, b) FQHE singularities in suspended graphene for the filling $\frac{1}{3}$ in a field of 2-12 T with a concentration of 10^{10}cm^{-2} and a mobility of 200000 cm^{2}V^{-1}s^{-1}. *Source: a) K. I. Bolotin, F. Ghahari, M. D. Shulman, H. L. Störmer, and P. Kim, "Observation of the fractional quantum Hall effect in graphene," Nature 462, p. 196, 2009. b) X. Du, I. Skachko, F. Duerr, A. Luican, and E. Y. Andrei, "Fractional quantum Hall effect and insulating phase of Dirac electrons in graphene," Nature 462, p. 192, 2009.*

FQHE in suspended graphene is observed at temperatures of approximately 10 K [Abanin *et al.* (2010)] or higher (up to 20 K) [Skachko *et al.* (2010)], which seems to be related to the stronger electric interaction in view of a lack of a dielectric substrate (with a relatively high dielectric constant in the case of semiconductors, ~ 10) in the case of suspended samples or, on the other hand, with very high cyclotron energy in graphene (i.e., a large energy gap between incompressible states). The fractional Hall effect in graphene is also considered in relation to the pseudospin structure in terms of symmetry SU(4) and SU(2) [Papić *et al.* (2010)].

In the papers [Du *et al.* (2009); Bolotin *et al.* (2009)] it has been demonstrated the competition between the FQHE state with the insulator state near the Dirac point accompanying a rapidly decreasing concentration (cf. Fig. 4.8).

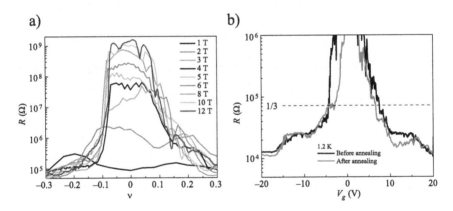

Fig. 4.8 a) The emergence of an insulator state that accompanies the increase in the strength of a magnetic field around the Dirac point, b) competition between the FQHE and the insulator state for the filling $-1/3$: annealing removes pollution – enhances mobility and provides conditions for the emergence of a *plateau* for FQHE. *Source: X. Du, I. Skachko, F. Duerr, A. Luican, and E. Y. Andrei, "Fractional quantum Hall effect and insulating phase of Dirac electrons in graphene," Nature 462, p. 192, 2009.*

From the perspective of cyclotron groups, the experimental results on the FQHE in graphene [Du *et al.* (2009); Bolotin *et al.* (2009); Abanin *et al.* (2010); Skachko *et al.* (2010); Papić *et al.* (2010)] seem to be in agreement with the expectations of the braid description. In the case of graphene, the specific band structure (one of a gapless semiconductor) with conical Dirac bands leads to the simultaneous participation (in the case of Dirac point) of both bands, of holes and of electrons, which when combined with the massless character of the Dirac fermions, manifests itself through an anomalous IQHE (corrected additionally in very strong fields as a result of eliminating spin and two-valley degeneration) [Novoselov *et al.* (2005); Zhang *et al.* (2005, 2006)]. Controlling the lateral gate voltage (within the range up to 20 V [Du *et al.* (2009)]) allows regulation of the density of the carriers under a constant magnetic field. One should therefore expect that at relatively small densities of carriers (electrons, or symmetrical holes at reverse voltage polarization), the cyclotron orbits will be too short to permit braid exchanges of particles under a sufficiently strong magnetic field, although weaker for smaller concentrations, and the experimental observations support this expectation [Du *et al.* (2009); Bolotin *et al.* (2009)]. For low concentrations, while closing on the Dirac point, one may expect that fields that are too strong would exceed the stability threshold of the IQHE state in competition with the Wigner crystal (assuming a similar character of this competition in the case of massless Dirac fermions in reference to traditional semiconductor 2D structures) and that corresponds to the emergence of the insulating state near the Dirac point in a strong magnetic field [Yannouleas *et al.* (2010)]. In the case of the hexagonal structure of graphene, electron (or hole) Wigner crystallization may exhibit interferences between the triangular crystal sublattices, and inclusion of the resonance (hopping) between these two sublattices may cause blurring of the sharp transition to the insulator state, which seems to be in agreement with the observations (Fig. 4.8).

It is worth noting that, in some correspondence to the honeycomb graphene structure, the topologically distinct quantum states in optical network systems filled with atoms (fermions or bosons) have recently been studied in depth [Stanescu *et al.* (2010)]. The implementation of such networks with hexagonal cells, *honeycomb*, or square cells, *checkerboard*, is practically possible with the use of interference from several appropriately oriented laser beams, and their specially adjusted modulation would even enable the rotation of elementary cells to be achieved (imitating a magnetic field) [Palmer and Jaksch (2006)]. This creates entirely new opportunities for studying strongly correlated planar systems (including boson systems, which are markedly different from electron systems), although the experimental progression has not been satisfactory thus far, particularly as a result of the inability to observe the transportation effects of neutral (as a rule) atoms trapped in optical networks. In theory, however, Hall states in a magnetic field are being studied in these systems [Palmer and Jaksch (2006)], as well as similar states without Landau levels [Sheng *et al.* (2011);

Zhang *et al.* (2011); Sørensen *et al.* (2005)]. The latter refer to Haldane's idea [Haldane (1988)], according to which, in the systems of flat periodic lattice (originally a honeycomb type), it is possible to introduce Berry topological curvature by modeling complex amplitudes of hopping to the nearest and next-nearest neighbors, which breaks the symmetry for time reversion and allows the modeling of topologically different quantum states, similar to states in a strong magnetic field [Sun *et al.* (2009)]. Such models are associated with the recently suggested topological insulators [Wang *et al.* (2010); Qi and Zhang (2010)] that are described by topological invariants similar to the ones previously proposed for Hall systems as well [Yakovenko (1990); Avron *et al.* (2003)]. These do not necessarily correspond to Hall states in semiconductor heterostructures or in graphene, but they may be related to the topological effects: e.g., for surface states [Cheng *et al.* (2010)]. However, in the case in which they refer to systems in a strong magnetic field (or one imitated by the rotation of elementary cells [Zhang *et al.* (2011)] by suitably chosen phases for hopping amplitudes [Sheng *et al.* (2011)]), we can expect similar relationships with cyclotron braid groups, as is the case for 2D electron systems in semiconductor quantum wells, as well as for graphene (the states of Wigner crystallization can also be considered in a similar way [Wu *et al.* (2007); Willett *et al.* (1988)]).

Chapter 5

Recent progress in FQHE field

5.1 The role of carrier mobility in triggering fractional quantum Hall effect in graphene

The FQHE is essentially a collective arrangement of electron states with primery role of the interaction. This is expressed in the form of the Laughlin wave function [Laughlin (1983a)], which turns out [Haldane (1983); Prange and Girvin (1990)] to be an exact ground state for a 2D system of charged particles in the presence of a magnetic field at the lowest Landau level (LLL) filling $1/p$ (p–odd), when the short-range part of Coulomb interaction is included. This has been demonstrated using so-called Haldane pseudopotentials, i.e., matrix elements of Coulomb interaction based on the relative angular momentum m of electron pairs. The short-range part of the interaction is limited by the $m = p - 2$ Haldane term for the p-th FQHE state, and it has been verified that the long-range tail with $m > p - 2$ does not significantly influence this state [Prange and Girvin (1990)]. Despite strong correlation effects, the basic physics of the FQHE was successfully grasped based on an effectively single-particle model of composite fermions (CFs) [Jain (1989)]. CFs are assumed to be residually interacting particles with $p - 1$ flux quanta of the auxiliary field fixed in some way to each particle. These composite particles gain the statistics required by the Laughlin function as a result of the Aharonov–Bohm effect, when particles interchange together with the flux tubes fixed to them. The CF concept is commonly accepted because of its appealing single-particle picture; further modeling of CFs by variants of the Chern–Simons field resulted in efficient calculation schemes, supported by good agreement with exact diagonalizations, especially inside the LLL [Heinonen (1998); Jain (2007)].

Recent experimental investigations of the FQHE in graphene [Bolotin et al. (2009); Du et al. (2009)] have shed, however, new light on this corre-lated state and seem to go beyond the explanative ability of CF treatment, which concentrates solely on the interaction. If one imagines CFs to be analogous to solid-state Landau quasiparticles dressed with the interaction, i.e., presuming that the local flux tubes are a result of the interaction solely [Heinonen (1998); Jain (2007)], one would lose important topological effects and encounter problems with new direct observations indicating that the carrier mobility plays a triggering role for the FQHE, as has been convinc-ingly demonstrated in suspended graphene samples. This demonstration [Du et al. (2009)] consists in the observation of the FQHE in the same sam-ple and under the same conditions but only after annealing, which enhances the carrier mobility without affecting the interaction. This agrees with the well-known requirements associated with sample high quality needed to ob-serve the FQHE in traditional semiconductor heterostructures, which has also been previously evidenced [Pfeiffer and West (2003)].

The topological arguments seem to be in agreement with recent obser-vations of suspended graphene, thereby highlighting the crucial role of the carrier mobility in addition to the interaction in FQHE formation. Simul-taneously, the topological conditioning of the FQHE Laughlin correlations is clarified.

A single-atom-thick layer in graphene creates a hexagonal 2D structure with two carbon atoms per unit cell. This results in a double triangular lattice—a hexagonal structure of carbon atoms arranged in a honeycomb pattern, as depicted in Fig. 4.5. The p orbitals, perpendicular to the plane, hybridize into type π of the band structure, well described in the approximation of strong coupling,

$$E_{\pm}(\mathbf{k}) = \pm t \sqrt{3 + f(\mathbf{k})} - t' f(\mathbf{k}), \qquad (5.1)$$

where $f(\mathbf{k}) = 2\cos(\sqrt{3}k_y a) + 4\cos(\frac{\sqrt{3}}{2}k_y a)\cos(\frac{3}{2}k_x a)$, $a \simeq 0.142$ nm is the distance between carbon atoms, $t = 2.7$ eV is the hopping energy between the nearest neighbors (between sublattices) and $t' = 0.2t$ is the hopping energy between next-nearest neighbors (inside the sublattices). This struc-ture is plotted in Fig. 4.5 (b). The valence band and the conduction band touch at points called K and K' at the border of a hexagonal Brillouin zone [Wallace (1947); Castro Neto et al. (2009)] in compliance with the above-mentioned relation for $t' = 0$ (i.e., when only nearest-neighbor hopping is included). Both bands that met at these points (non-gap semiconductor) are locally an ideal conical shape, which means that the relation between energy and momentum (counted from points of contact) is linear and the appropriate band Hamiltonian is formally equivalent to the model for relativistic fermions with zero rest mass ($E = \pm\sqrt{m_0^2 v_F^4 + p^2 v_F^2}$; in this equation, $m_0 = 0$) which are described by the Dirac equation, where the velocity of light is replaced by the Fermi velocity, $v_F \simeq c/300$ [Castro Neto *et al.* (2009); Yang (2007)]. Therefore, for Dirac electrons (with energy linear in momentum, such as close to K and K' points), the equation describing local quantum dynamics can be expressed as follows, $-iv_F\vec{\sigma} \cdot \nabla\Psi(\mathbf{r}) = E\Psi(\mathbf{r})$, where the Pauli matrix vector corresponds to the pseudospin structure related to two sublattices [Castro Neto *et al.* (2009); Geim and MacDonald (2007)] (wave functions are spinors in this structure). As mentioned in the subsection 4.9, the zero mass of the Dirac fermions leads to numerous consequences and electron anomalies in the properties of graphene [Castro Neto *et al.* (2009); Geim and MacDonald (2007); Novoselov *et al.* (2005); Zhang *et al.* (2005)]. For linear dependence of energy and momentum, the momentum uncertainty also leads to an energy uncertainty (contrary to non-relativistic case), which mixes in the time-evolution particle states with hole (anti-particle) states. The Klein paradox, which refers to the ideal tunneling of Dirac particles through rectangular potential barriers, leads to high charge-carrier mobility in graphene, which has been experimentally observed, in agreement with theoretical predictions [Castro Neto *et al.* (2009); Novoselov *et al.* (2005); Zhang *et al.* (2005); Yang (2007)].

The zero mass of Dirac electrons also changes the scaling of cyclotron energy in comparison to the non-relativistic case. In the relativistic case, $\omega_c \sim B^{1/2}$, and not $\sim B$, as in the case of non-relativistic particles. The value of this energy, $\hbar\omega_c$, is two orders of magnitude larger than that corresponding to classical materials due to zero mass at the Dirac point and reaches a very high value of order of 1000 K (for 10 T field). This allows for the observation of the IQHE in graphene even at room temperature [Novoselov *et al.* (2005); Zhang *et al.* (2005)]. An anomalous IQHE is observed for filling fractions $\nu = \pm 4(n + 1/2)$, i.e., for $\pm 2, \pm 6, \pm 10, \ldots$ as illustrated in Fig. 4.6. The distinction with respect to the IQHE in semiconductor 2DEG is caused by the location of the zero Landau level at the Dirac point, i.e., for zero energy. Moreover, in the formula $\nu = \pm 4(n+1/2)$, $\pm$ corresponds to particles and holes, respectively, the value 4 results from pseudospin/valley degeneracy and $1/2$ is associated with the Berry phase for Dirac particles [Castro Neto *et al.* (2009); Yang (2007); Geim and MacDonald (2007); Novoselov *et al.* (2005); Zhang *et al.* (2005); MacClure (1956)].

Despite using very strong magnetic fields (up to 45 T), FQHE was not detected in graphene samples deposited on a SiO_2 substrate [Zhang et al. (2006)]. Ref. [Zhang et al. (2006)] noted, however, the emergence of additional plateaus of the IQHE for the fillings $\nu = 0, \pm 1, \pm 4$, indicating the elimination of spin and pseudospin (related to sublattices) degeneration as a result of acquiring mass by Dirac fermions [Zhang et al. (2006)].

After mastering the novel technique of so-called suspended ultrasmall graphene scrapings to produce samples of extremely purity and with a high mobility of carriers (beyond 200000 $cm^2V^{-1}s^{-1}$—note that high mobility is also necessary to observe the FQHE in the case of semiconductor 2D heterostructures, in which even higher mobilities are reached, i.e., millions of $cm^2V^{-1}s^{-1}$ [Pfeiffer and West (2003)])—was it possible to observe the FQHE in graphene at fillings $\nu = 1/3$ and $-1/3$ (the latter for holes, with opposite polarization of the gate voltage, which determines the position of the Fermi level, either in the conduction band or in the valence band) [Du et al. (2009); Bolotin et al. (2009)]. Both papers report the observation of the FQHE in graphene for moderately strong magnetic fields. In the paper [Bolotin et al. (2009)] it is reported that the FQHE was observed at $12-14$ T and an electron concentration of $10^{11}/cm^2$ (a mobility of 250000 $cm^2V^{-1}s^{-1}$), and in the paper [Du et al. (2009)], the FQHE was reported that was observed in a weaker field with a magnitude of $2-12$ T but at a concentration level that was smaller by one order of magnitude $10^{10}cm^{-2}$ (a mobility of 200000 $cm^2V^{-1}s^{-1}$)).

The FQHE in suspended graphene has been observed at temperatures around 10 K [Abanin et al. (2010)] and even higher (up to 20 K) [Skachko et al. (2010)]. The critical temperature enhancement is likely related to the stronger electric interaction caused by a lack, in the case of suspended samples, of a dielectric substrate with a relatively high dielectric constant in the case of semiconductors and the very high cyclotron energy in graphene. The higher stability of the FQHE in suspended graphene is also linked to higher mobilities. One can expect that in the ideal 2D system of suspended graphene, the interaction between carriers and phonons is suppressed in comparison to the 3D case of graphene on a solid substrate, which leads to mobility enhancement in suspended samples, which is important for the organization of FQHE state. The role of carrier mobility has been demonstrated in [Du et al. (2009)], where it has been shown that the FQHE occurs in the same sample, which is originally insulating under the same conditions but after annealing process enhancing carrier mobility due to a reduction in impurity content, as presented in Fig. 4.8 (b). This observation directly demonstrates the triggering role of carrier mobility in FQHE state arrangement.

The competition between the FQHE state and the insulator state near the Dirac point, corresponding to a rapidly decreasing carrier concentration (and thus reducing the interaction role at larger separations of carriers), has also been demonstrated [Du *et al.* (2009); Bolotin *et al.* (2009)]—Fig. 4.8.

Very recent experimental progress has also allowed for the observation of the FQHE in graphene on a $h - BN$ (hexagonal boron nitride) substrates in large magnetic fields on the order of 40 T (remarkably, FQHE features were noticed at higher Landau levels, which is a record in this regard) [Dean *et al.* (2011)].

The peculiar FQHE behavior in graphene, with the direct manifestation of the conditioning of this collective state by the carrier mobility and not only by the interaction, requires non-local topology-type arguments beyond only a single-particle effective CF model with flux tubes attached to particles.

The relevant topological theory of 2D charged multiparticle systems under strong magnetic fields was developed in the framework of the cyclotron braid group approach [Jacak *et al.* (2009a, 2010b)]. The main ideas of this formulation are as follows:

- For fractional fillings of the LLL, classical cyclotron orbits are too short for particle exchanges (as in the 2D case, long-range helical motion is impossible).
- The exchanges are necessary to create a collective state, such as the FQHE; thus, the cyclotron radius must be enhanced (in the model of Jain's CFs [Jain (1989)] this enhancement is achieved by the artificial addition of flux tubes directed opposite to the external field; in the model of Read's vortices [Read (1994)], a reduction in the cyclotron radius is attained by the depletion of the local charge density—both of these tricks are, however, phenomenological of effective character and not supported in topological terms).
- The topologically justified way to enhance the cyclotron radius is through the use of multi-looped cyclotron trajectories related to multi-looped braids that describe elementary particle exchanges in terms of braid groups (the resulting cyclotron braid subgroup is generated by σ_i^p, $i = 1, ..., N$, where σ_i are generators of the full braid group); in 2D, all loops of a multi-looped trajectory must share together the same total external magnetic field flux, in con-

trast to the 3D case, and this is the reason for enhancing all loop dimensions exactly fitting the particle separation at LLL fillings $1/p$ (p-odd).

- In accordance with the rules of path integration for non-simply-connected configuration spaces [Laidlaw and DeWitt (1971)], one-dimensional unitary representations (1DURs) of a corresponding braid group define the statistics of the system—in the case of multi-looped braids, which are naturally assembled into a cyclotron sub-group, one arrives in this way at the statistical properties required by the Laughlin correlations (these 1DURs are $\sigma_i^p \to e^{ip\alpha}$, $\alpha \in [0, 2\pi)$, CFs correspond to $\alpha = \pi$).

- The interaction is important for properly determining the cyclotron braid structure because its short-range part prevents the particles from approaching one another closer than the distance given by the density.

All of these ideas are summarized in Fig. 5.1 (for details cf. subsections 4.1–4.6).

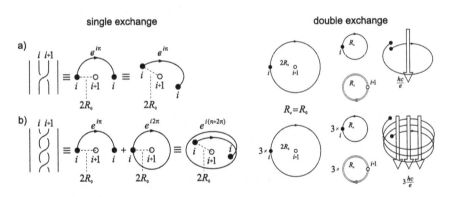

Fig. 5.1 The braid generator corresponds to the single exchange of particles (left), the cyclotron orbit corresponds to the double exchange (right) for $\nu = 1$ when a single-looped cyclotron trajectory reaches neighboring particles, $R_c = R_o$ (a), and for $\nu = \frac{1}{3}$, with two additional loops needed for $R_c = R_o$ (b).

This completes a topological derivation of FQHE statistics without needing to model it by fictitious auxiliary elements such as flux tubes; moreover, it proves that CFs are not quasiparticles dressed with interaction (as familiar for condensed-matter Landau quasiparticles) or complexes with local fluxes but are rightful quantum 2D particles assigned with Laughlin statistics determined by the 1DURs of the appropriate cyclotron braid subgroup.

Nevertheless, because the model of CFs with attached rigid flux quanta works so well (as evidenced by the exact diagonalizations [Heinonen (1998); Jain (2007)]), the multi-looped classical braid structure must be repeated by quasiclassical wave-packet trajectories (and then with quantized fluxes, which was not, however, a rule for classical trajectories). Note that flux quantization is a quasiclassical property because it requires a trajectory definition (the carrier mobility also has a similar quasiclassical background).

The wave packets corresponding to the quasiclassical dynamics are related to the collective character of a multiparticle system. For the LLL of a non-interacting system, the group velocity of any packet is zero due to the degeneracy of states. Interaction removes, however, this degeneracy and provides packet dynamics. The collective movement minimizes kinetic energy, whereas the interaction favors localization (which causes a related increase in kinetic energy). Therefore, the collective dynamics prefers the quasi-classical movement of packets along periodic closed trajectories (as a rule in the presence of a magnetic field [Eliutin and Krivchenkov (1976)]), which then must, however, embrace quantized external magnetic field fluxes. This is the role of collectivization associated with the energy preference of wave packets traversing closed trajectories in correspondence with the classical cyclotron description, including the multi-looped braid picture [Jacak *et al.* (2010b)].

This description appears to be in accordance with FQHE observations in graphene (described above), which are found at a low carrier density and therefore accompany their dilution and the resulting reduction in interaction. Thus, the interaction is not the sole factor initiating the FQHE, as was previously expected in view of the standard model of composite fermions, if one treated the dressing of fermions with localized flux tubes as a result of just only the interaction itself [Heinonen (1998); Jain (2007)]. Carrier mobility refers to semiclassical wave-packet dynamics in terms of the drift velocity in an electric field and the classical Hall effect and reflects various channels of scattering phenomena beyond the simple model that includes only Coulomb interaction to free particles in a magnetic field. However, topology arguments in the 2D case strongly prefer high mobility, which is required for real wave packets to traverse multi-looped trajectories. Semiclassical wave packets, even in the presence of interaction and scattering, manifest periodic dynamics [Eliutin and Krivchenkov (1976)], and in the case of the multi-looped trajectory structure with enhanced radii, higher mobility is required, as has been experimentally demonstrated.

From the cyclotron group point of view, the experimental results for the FQHE in graphene [Du *et al.* (2009); Bolotin *et al.* (2009)] seem to be compliant with the expectations of the braid description. In the case of suspended graphene, controlling the lateral gate voltage (within the range up to 20 V [Du *et al.* (2009)]) allows for the regulation of the density of carriers at a constant magnetic field. One should therefore expect that at relatively small densities of carriers (electrons, or symmetrical holes at reverse voltage polarization), the cyclotron orbits will be too short to prevent braid exchanges of particles in a sufficiently strong magnetic field—although weaker for smaller concentrations; indeed, experimental observations have supported exactly this prediction [Du *et al.* (2009); Bolotin *et al.* (2009)]. At low concentration, while close to the Dirac point, one may expect that excessively strong fields would exceed the stability threshold of the FQHE state in competition with the Wigner crystal (taking into account a specific character of this competition in the case of massless Dirac fermions in comparison to traditional semiconductor 2D structures [Dahal *et al.* (2006)], resulting in the absence of a Wigner transition without a magnetic field); this corresponds to the emergence of the insulating state near the Dirac point in a strong magnetic field. In the case of the hexagonal structure of graphene, electron (or hole) Wigner crystallization may exhibit interference between the triangular crystal sublattices, and the inclusion of the resonance (hopping) between these two sublattices may blur the sharp transition to the insulator state, which seems to agree with observations (Fig. 4.8).

Some specific FQHE features in graphene may also be linked to the spin-valley SU(4) symmetry of Dirac carriers [Dean *et al.* (2011); Goerbig (2011)]. In order to account for this four-fold structure, one can use the multicomponent model of FQHE wave function suited to SU(4) symmetry in the form of the Halperin wave function [Halperin (1983); Goerbig and Regnault (2007)]. The Halperin wave function is a generalization to SU(2) of the Laughlin function (with magnetic length, $l = 1$),

$$\Psi_m^L(\{z_k\}) = \prod_{k<l}^{N}(z_k - z_l)^m e^{-\sum_k^N |z_k|^2/4}, \qquad (5.2)$$

to two-spin-component system in the following form,

$$\Psi_{m\downarrow}^L(\{z_{k\downarrow}^\downarrow\})\Psi_{m\uparrow}^L(\{z_{k\uparrow}^\uparrow\}), \qquad (5.3)$$

with possible inter-component correlation factor,

$$\prod_{k_\downarrow}^{N_\downarrow}\prod_{k_\uparrow}^{N_\uparrow}(z_{k_\downarrow}^\downarrow - z_{k_\uparrow}^\uparrow)^n. \tag{5.4}$$

where the exponent n may be even or odd, while m, $m \downarrow$ and $m \uparrow$ are odd.

The two-component Halperin function can be easy generalized to the SU(N) case [Goerbig and Regnault (2007)] and in the case of graphene with the spin-valley SU(4) symmetry was introduced in the following form [Papić *et al.* (2009a)],

$$\begin{aligned}\Psi_{m_1,\ldots,m_4,n_{i,j}}^{SU(4)} &= \prod_j^4 \prod_{k_j < l_j}^{N_j} (z_{k_j}^j - z_{l_j}^j)^m e^{-\sum_{j=1}^4 \sum_{k_j=1}^{N_j} |z_{k_j}^j|^2/4} \\ &\times \prod_{i<j}^4 \prod_{k_i}^{N_i} \prod_{k_j}^{N_j} (z_{k_i}^i - z_{k_j}^j)_{i,j}^n,\end{aligned} \tag{5.5}$$

m_j must be odd integers, whereas n_{ij} may also be even integers. Because the components of this function assume the form of a Jastrow polynomial, a topological interpretation similar to that in the single-component electron liquid case is required.

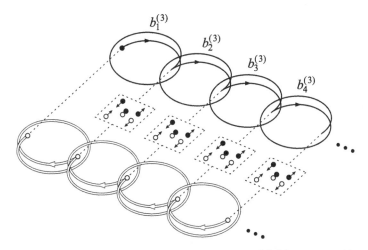

Fig. 5.2 The multibraid of the form of generator product $b_1^{(p)} b_2^{(p)}...$, where $b_i^{(p)}$ are generators of the cyclotron braid subgroup (in the figure $p = 3$); this multibraid corresponds to the arrangement of an edge-to-edge cyclotron hopping pathway, due to collisions between every second particle; completing of $\frac{p-1}{2}$-looped braids (generators $b_i^{(p)}$) is required before colliding between the every second particle; these all require sufficiently longer mean free path due to additional $\frac{p-1}{2}$ loops for each neighbors exchange.

The carrier mobility in suspended graphene reaches $250\,000$ cm^{2}V^{-1}s^{-1}. It is, however, lower than the record value for 2D semiconductor structures, which is ca 30×10^6 cm^{2}V^{-1}s^{-1} [Pfeiffer and West (2003)]. However, taking into account the fact that the carrier concentration in graphene can be lower than that in semiconductor heterostructures [Pfeiffer and West (2003); Bolotin *et al.* (2008)], the corresponding mean free path in both cases well exceeds the sample dimensions (of μm order in both cases, because the mobility is proportional to the concentration and to the mean free path of carriers). In the cyclotron braid approach, one can argue that this corresponds to the manifestation of multibraids (i.e., not only of single generators but also of their products), which in a magnetic field allow particles (represented by quasiclassical wave packets repeating braids exchanges) to traverse edge-to-edge pathways through a series of cyclotronic exchanges and collisions in an equidistantly distributed particle network prevented by Coulomb interaction. This emphasizes the central role of Coulomb repulsion in FQHE formation but also that of the sufficiently long mean free path of carriers and the related mobility enhancement. Taking into account the fact that cyclotronic orbits must be p-looped to reach neighboring particles at the LLL filling $\frac{1}{p}$ and to allow the series of exchanges needed to edge-to-edge pathways, the resulting mean free path should be p times longer than the sample dimensions, which seems to fit experimental observations in both graphene [Bolotin *et al.* (2008)] and semiconductor heterostructures [Pfeiffer and West (2003)]. This cyclotron multi-looped edge-to-edge hopping is schematically presented in Fig. 5.2: neighboring particles simultaneously exchange positions, and when the cyclotron orbit dimension fits the particle separation, the edge-to-edge hopping is arranged due to collisions between every second particle. When the LL filling is equal to $\frac{1}{p}$, (p—odd integer), to exchange neighboring particles, the $\frac{p-1}{2}$-lopped braid orbits are required [Jacak *et al.* (2009a, 2010b)]. These particles (called composite fermions) cannot collide with every second particle before completing an exchange with the nearest neighbors (it needs to complete all loops of the cyclotron braid corresponding to the cyclotron braid generators $b_i^{(p)} = \sigma_i^p$, σ_i—generators of the full braid group). This requires a p times longer mean free path (and p times greater related mobility) compared to the single-looped exchanges of ordinary electrons.

Although the real dynamics of quasiclassical wave packets is beyond the framework of a simplified 2D multiparticle charged system in a magnetic field, some general qualitative conclusions regarding the topological character can be drawn. In the case of the fractional filling of the LLL, when classical cyclotron orbits are too short for braid exchanges, multi-looped

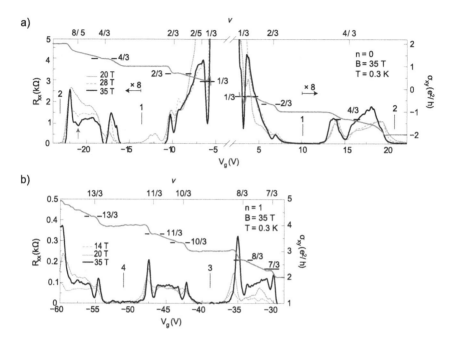

Fig. 5.3 Fractional quantum Hall effect in graphene, a,b. Magnetoresistance (left axis) and Hall conductivity (right axis) in the $n = 0$ and $n = 1$ Landau levels at $B = 35$ T and temperature ~ 0.3 K; filling factors for IQHE plateaus, $\nu = \pm 4(n + 1/2)$ are linked to Landau index $n = 0, 1, 2, ...$ due to spin-valley four-fold degeneracy (SU(4) structure) and due to the Barry phase (1/2) offset for Dirac particles; thus, LL $n = 0$ corresponds to the filling factor range $0 < \nu < 2$ and $n = 1$ to $2 < \nu < 6$. *Source: C. R. Dean, A. F. Young, P. Cadden-Zimansky, L. Wang, H. Ren, K. Watanabe, T. Taniguchi, P. Kim, J. Hone and K. L. Shepard, "Multicomponent fractional quantum Hall effect in graphene", Nature Physics 7, p. 693, 2011.*

trajectories occur. Through the linkage of this classical cyclotron dynamics with quasiclassical wave-packet trajectories, supported by the success of the CF model with rigid flux quanta attached to particles, one can also expect multi-looped structures of real quasiclassical wave-packet dynamics for which flux quantization is held obligatorily (for classical braid orbits it was not a rule). With larger radii, the multi-looped orbits strongly favor higher carrier mobility, which has been confirmed experimentally in suspended graphene. This suggests that the mobility is of primary significance for FQHE formation in competition with localized electron states such as insulating Wigner-crystal-type states. This picture seems to also agree with the already-predicted [Jain (2007)] destabilization of Laughlin states in semiconductor heterostructures at $p > 9$ in favor of a Wigner crystal when visibly too many loops begin to be energetically unfavorable. Experiments with graphene have indicated that not only interaction influences the stability of FQHE states, which agrees with the topological explanation of CFs properties, thereby demonstrating that the previous concept of flux tubes generated by the interaction itself is insufficient in this regard.

Fig. 5.3 shows the FQHE features for $n = 0, 1$ LLs, which correspond to the range of fillings $\nu \in [0, 6]$ according to the four-fold degeneracy and the Berry phase shift for Dirac particles, which gives the relation between Landau index n and filling factor ν,

$$\nu = \pm 4(n + \frac{1}{2}). \tag{5.6}$$

At these fillings, the IQHE is observed in graphene; thus, for $\nu = \pm 2, \pm 6$ correspond to $n = 0, 1$. In high-quality samples, the IQHE is additionally observed for all integers ν (cf. Figs. 5.3 and 5.4, where the IQHE plateaus are also marked, up to 4) due to the ferromagnetic polarization of the SU(4) structure [Bolotin *et al.* (2009); Du *et al.* (2009)] in high magnetic fields. In low fields of the order of $1 - 2$ T, the Shubnikov-de Haas oscillations are clearly visible, as shown in the inset in Fig. 5.4.

5.2 Development of Hall-type experiment in conventional semiconductor materials

Since the early observations of the FQHE in 2D semiconductor heterostructures [Tsui *et al.* (1982)], continuous improvements in sample manufacturing and quality have allowed for considerable enhancements in carrier mobility over a wide concentration range, as described in, e.g., Ref. [Pfeiffer and West (2003)]. The highest reported ([Pfeiffer and West (2003)]) mobility is equal to 31×10^6 cm^{2}V^{-1}s^{-1} for concentrations in the range $10^9 - 10^{11}$ 1/cm^2. The previous record, reported in [English *et al.* (1987)], was ca 6 times lower, 5×10^6 cm^{2}V^{-1}s^{-1}, while in the original discovery sample [Tsui *et al.* (1982)] the mobility was only 9×10^4 cm^{2}V^{-1}s^{-1}. Remarkably, the mobility μ scales as $\mu \sim n^{0.7}$ with density n in the above-mentioned window $10^9 - 10^{11}$ 1/cm^2 [Pfeiffer and West (2003)].

The progress made in techniques of sample manufacturing with high quality has allowed for new experimental possibilities. Only at sufficiently high mobilities can the additional effects of Shubnikov-de Haas oscillations and magnetoresistance vanishing in low magnetic fields be observed, although in the presence of GHz microwaves (94 GHz, and a mobility 3×10^6

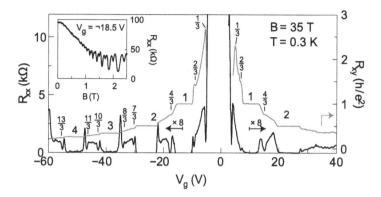

Fig. 5.4 Integer and fractional quantum Hall effect features in graphene; magnetoresistance (left axis) and Hall resistance (right axis) in the $n = 0$ and $n = 1$ Landau levels at $B = 35$ T and temperature ~ 0.3 K; all integer filling factors for IQHE plateaus are indicated; inset demonstrates Shubnikov-de Haas oscillations in a magnetic field on the order of 1 T. *Source: C. R. Dean, A. F. Young, P. Cadden-Zimansky, L. Wang, H. Ren, K. Watanabe, T. Taniguchi, P. Kim, J. Hone and K. L. Shepard, "Multicomponent fractional quantum Hall effect in graphene", Nature Physics 7, p. 693, 2011.*

cm^{2}V^{-1}s^{-1} of 2DEG) [Zudov *et al.* (2001)]. By applying 103 GHz microwaves to a 15×10^6 cm^{2}V^{-1}s^{-1} high-mobility 2DEG system, these new oscillations in R_{xx} were observed to be too in low magnetic fields (apart the ordinary Shubnikov-de Haas oscillations), although not those in R_{xy}—cf. Fig. 5.5, after [Mani *et al.* (2002)], also confirmed at 27×10^6 cm^{2}V^{-1}s^{-1} by the application of lower-frequency, 57-GHz radiation, as presented in Fig. 5.6 after [Zudov *et al.* (2003)].

Similarly, at such large mobilities, some new observations of FQHE features near $\nu = \frac{3}{8}$ (at $\frac{5}{13}$ and $\frac{4}{11}$) have been reported—cf. Fig. 2.1. As these LLL filling factors are outside the main hierarchy line, thus they do not align with the traditional CF picture; the idea of CF interaction was heuristically developed [Jain (2007)] to explain these observations. Note that in the cyclotron braid approach, in compliance with subsection 4.6, it could be, however, explained without such an artificial assumption.

The next interesting observation linked to sample quality is the FQHE at $\nu = \frac{5}{2}$, which reveals the re-entrant insulating phases between neighboring IQHE plateaus sensitive to high mobility values [Eisenstein *et al.* (2002)], as is depicted in Fig. 5.7.

Note, however, that recent observations of the FQHE in high-mobility graphene samples (Fig. 5.3) reaches a record filling factor of 4, which belongs to the $n = 1$ LL according to the formula $\nu = \pm 4(n + 1/2)$ (here $\pm$

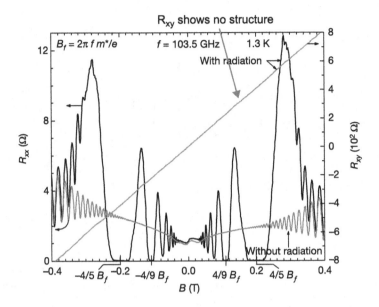

Fig. 5.5 Microwave-induced magneto-resistance oscillations and zero-resistance states under 103 GHz microwave photo-excitation in the high mobility GaAs/AlGaAs 2D electron system. *Source: R. G. Mani, J .H. Smet, K. v.Klitzing, V. Narayanamurti, W. B. Johnson and V. Umansky, "Zero-resistance states induced by electromagnetic-wave excitation in GaAs/AlGaAs heterostructures," Nature 420, p. 646, 2002.*

refers to particles and holes, 4 reflects the four-fold spin-valley degeneracy, i.e., SU(4) symmetry and 1/2 is due to Berry's phase for Dirac particles) [Dean *et al.* (2011)]. Energy gaps protecting incompressible FQHE states are larger in graphene than in traditional semiconductor materials, reaching a magnitude of the order of 16 K (at $\nu = \frac{4}{3}$ and $B = 35$ T), cf. Fig. 5.8, which is referred to as the Dirac massless character of carriers. In conventional semiconductor heterostructures, the corresponding gaps are much lower and the observed FQHE stability with temperature is much more fragile.

The triggering role of high carrier mobility in the formation of FQHE collective states in traditional semiconductor material also indicates that quasiclassical wave-packet dynamics plays an important role, which can be observed using the cyclotron braid approach and is rather beyond the scope of the local-type effective CF model with auxiliary flux tubes fixed to particles.

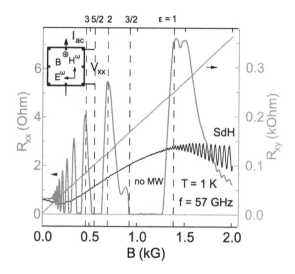

Fig. 5.6 Zeros in R_{xx} in the presence of 57-GHz microwaves and mobility 27×10^6 cm^{2}V^{-1}s^{-1} 2DEG. *Source: M. A. Zudov, R. R. Du, L. N. Pfeiffer and K. W. West, "Evidence for a new dissipationless effect in 2D electronic transport," Phys. Rev. Lett. 90, p. 046 807, 2003.*

5.3 Topological insulators—new state of condensed matter

Investigation of IQHE and then of FQHE in 2D charged systems opened a broad area of topologically conditioned effects [Avron *et al.* (2003)]. In this regard, the current understanding of the IQHE treated in single-particle topology terms has been considerably developed, particularly with respect to the novel view on standard band structure [Qi and Zhang (2011)]. This is based on observation that the corresponding multi-particle IQHE states protected by Landau quantization gaps are not connected with symmetry breaking as many other condensed matter phases in scenario of ordinary phase transitions, but rather with topological invariants associated to a particular geometry and matter organization [Qi and Zhang (2010); Hasan and Kane (2010)]. These invariants are becoming increasingly better recognized in terms of homotopy groups related to specially defined multidimensional transformations of physically conditioned objects such as Green functions and their derivatives [Wang *et al.* (2010); Qi (2011)], which were previously developed to describe topology of textures in multicomponent condensed-matter states with rich matrix order parameters, including superfluid He^3 and liquid crystals [Mermin (1979)]. The role of various factors protecting

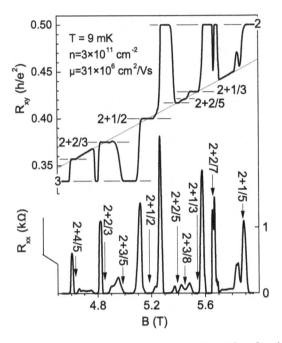

Fig. 5.7   FQHE in second LL in high-mobility sample, 31×10^6 cm²V⁻¹s⁻¹ 2DEG, re-entrant insulating behavior near both 5/2 and 7/2, where, as the magnetic field is slowly increased, R_{xy} oscillates between the expected FQHE plateaus and the nearby IQHE plateaus. *Source: J. S. Xia, W. Pan, C. L. Vincente, E. D. Adams, N. S. Sullivan, H. L. Störmer, D. C. Tsui, L. N. Pfeiffer, K. W. Baldwin and K. W. West, "Electron correlation in the second Landau level; a competition between many, nearly degenerate quantum phases", Phys. Rev. Lett. 93, p. 176809, 2004.*

gaps separating flat bands (almost degenerated, as LLs at the interaction presence, which were massively degenerated in the absence of the inter-action) are of particular interest in view of the role of the magnetic field breaking time reversal or other effects like spin-orbit interaction or special type (time-reversal breaking) traversing-arrangements around the closed loop inside an elementary cell with complex hopping constants (the latter is particularly well defined in 2D, though some closed loops in 3D would be also selected [Hasan and Moore (2011)]). The generalization of the familiar in mathematics Chern invariants[1] [Prodan (2011)] are developing in order

---

[1]With the help of this topological Chern invariants associated to space mappings it is possible to characterize and classify geometrically nonequivalent objects which cannot be continuously (homotopically) transformed one into another one, as e.g. the sphere and the torus or manifolds with greater number of holes.

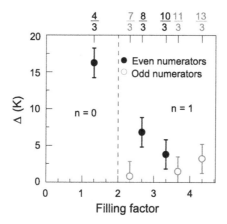

Fig. 5.8 FQHE energy gaps in zeroth and first LLs in high-mobility sample of graphene on a $h - BN$ (hexagonal boron nitride) substrate at $B = 35$ T and temperature $T \sim$ 0.3 K. *Source: C. R. Dean, A. F. Young, P. Cadden-Zimansky, L. Wang, H. Ren, K. Watanabe, T. Taniguchi, P. Kim, J. Hone and K. L. Shepard, "Multicomponent fractional quantum Hall effect in graphene", Nature Physics 7, p. 693, 2011.*

to grasp the essential topology of various multiparticle structures [Qi and Zhang (2011); Hasan and Kane (2010); Prodan (2011)]. The mappings of the Brillouin zone into state related objects can be in that manner classified by disjoint classes corresponding to topologically nonequivalent band orga- nizations protected by energy gaps conditioned by various physical factors and leading to distinct incompressible states (if gaps are related to system volume, as is typical in the 2D case, where in relation to quantum Hall effects, the area of the system enters via the external field total flux) in analogy to their prototype in the form of IQHE.

The distinctive character of 2D space is linked here to the magnetic field flux quantization, which is, as a rule, the quasiclassical property of trajectory loops as two-dimensional objects. The more general formulation of Chern-like topological invariants opens, however, a way to treating non- trivial situations of analogous topology in higher dimensions, despite the fact that spaces with more than 2 dimensions are inherently much more modest in terms of topology compared to the 2D case. The most important achievement in this regard was the determination of Z_2 invariants [Prodan (2011); Hasan and Kane (2010); Qi and Zhang (2011)] by the association of the appropriate invariant with five-dimensional mapping (of five dimen- sional k space to nonsingular Green functions in covering 4D+1 model), with the nontrivial π_5 homotopy group being the Z group [Wang *et al.*

(2010)]. In 3D+1 space, this invariant for an *interacting* system attains the following form when projected to lower dimensions,

$$P_3 = \frac{\pi}{6} \int_0^1 du \int d^4k/(2\pi)^4 Tr\epsilon^{\mu\nu\rho\sigma}$$
$$\times [G\partial_\mu G^{-1} G\partial_\nu G^{-1} G\partial_\rho G^{-1} G\partial_\sigma G^{-1} G\partial_u G^{-1}], \tag{5.7}$$

where $G(k, u)$ ($k = [\mathbf{k}, k_0]$ and u is an extension of the variable set distinguishing the sample edge from its inside) is the single-particle Green function for an interacting system; the momentum is integrated over the Brillouin zone. This topological order parameter resolves itself to Z_2 invariant for time-inversion symmetrical 2D+1 systems (and similarly in 3D+1 case) [Wang *et al.* (2010)]. This two-element homotopy group, which discriminates between different types of collective-state arrangements expressed via the invariant defined on the related Brillouin zone, gives rise to the conditioning of two distinct gapped states: the ordinary insulator and the topological insulator, which are associated with the 0 and 1 elements of the Z_2 group, respectively. Despite the local similarity between these two gapped states, the global arrangement of the band, noticeable only nonlocally (over the entire Brillouin zone), induces different overall behavior of the system. In the case of a topological insulator, one deals with insulating state inside the sample, whereas with the conducting nondissipative state on the sample edge, which is protected topologically, what there is, however, no case for an ordinary insulating state. This surprising phenomenon was confirmed experimentally and stimulated great interest.

From the point of view of band organization, topological insulators must be characterized by flat bands that meet in summits of locally cone-shaped valleys resembling Dirac points in graphene. This changes the topology and allows Chern-type invariants to attain nonzero values, which are attributed to the emergence of different global states. Spin degrees of freedom are of high significance with regard to topological arrangement, and related spin-type topological insulators are referred as to spin IQHE.

5.3.1   *Chern topological insulators*

The widely accepted definition of a new condensed-matter state called a topological insulator resolves to the robust metallic character of the edge or surface states and extended bulk insulating states that are also robust against disorder. This is amazing behavior, especially in two-dimensional models; when an edge or a surface is cut in a topological insulator sample, the emerging edge states are apparently connected to these bulk states.

These metallic edge states can be regarded as the terminating at the boundaries of the extended bulk states. This means that the bulk and the edge properties of topological insulators are mutually dependent and equally important. A similar situation occurs in IQHE, for which the emergence of dissipationless charge currents flowing around the edges of any finite IQHE sample is characteristic. Note, however, that the IQHE has been observed only in the presence of an externally applied magnetic field.

In 1988, Haldane presented a model of a condensed-matter phase that exhibits the IQHE without the need of a macroscopic magnetic field [Haldane (1988)]. The general idea of this effect can be sketched by writing a model Hamiltonian for a system of spinless particles occupying a honeycomb-type planar lattice with one state $|n>$ per site,

$$\hat{H} = \sum_{<n,m>} |n><m| + \sum_{\ll n,m \gg} [\xi_n |n><m| + h.c.] , \qquad (5.8)$$

where $< n, m >$ indicates the summation over nearest neighbors and the symbol $\ll n, m \gg$ indicates that the summation also includes next-nearest neighbors; the hopping factor $\xi_n = 0.5(t+i\eta)\alpha_n$ is assumed artificially as to be a complex number, and $\alpha_n = \pm 1$, depending on how n is positioned in the unit cell (equivalently, an isospin can be introduced here), as shown in Fig. 5.9 (note that in this lattice, the next-neighbor hopping always links sites with the same α). This Hamiltonian has two parameters t, η. The essence of the topological effect is linked to the imaginary contribution to the hopping factor given by η. The band structure corresponding to this Hamiltonian depends on parameters t, η and exhibits nontrivial topological properties (expressed by the Chern number C) for values t, η, as shown in the phase diagram in Fig. 5.10.

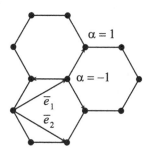

Fig. 5.9   The honeycomb structure (similar to that in graphene) for the Haldane model [Haldane (1988)]; e_1, e_2 are Bravais lattice vectors; nonequivalent site positions in the unit cell are indicated by $\alpha = \pm 1$.

The difference between a quantum Hall state and an ordinary insulator is a matter of topology [Thouless *et al.* (1982)]. According to Bloch theorem, the band structure corresponds to mapping from the momentum **k** defined on a torus of the Brillouin zone (in 2D case) to the Bloch Hamiltonian $H(\mathbf{k})$. Gapped band structures can be classified topologically in terms of equivalence classes of this mapping that cannot be continuously deformed into one another without closing the energy gap. Such classes are assigned by integer topological invariants which are called Chern numbers. The Chern numbers were introduced in the theory of fiber bundles [Nakahara (1990)]. They can be expressed physically by means of the Berry phase [Berry (1984)] associated with the Bloch wave functions $|u_m(\mathbf{k}) >$. The Bloch function acquires a Berry phase given by the line integral of $\mathbf{A}_m = i < u_m|\nabla_k|u_m >$, when **k** traverses a closed loop. Due to Stokes theorem, this line integral can be expressed by a surface integral of the Berry flux $\mathbf{F}_m = \nabla \times \mathbf{A}_m$. The Chern invariant can be defined as the total Berry flux passing the Brillouin zone,

$$C = \frac{1}{2\pi} \int d^2 k \mathbf{F}_m \in Z, \qquad (5.9)$$

C is an integer for reasons connected to unambiguity requirements, similar to the quantization of the Dirac magnetic monopole. The Chern number, C, is a topological invariant in the sense that it cannot change when the Hamiltonian varies smoothly. This manifests itself in the quantization of conductivity in the IQHE [Thouless *et al.* (1982)]. Here, one can use a simple instructive pictorial analogy. Instead of maps from the Brillouin zone to a Hilbert space, one can consider maps from two to three dimensions. Such mappings describe surfaces. 2D surfaces can be classified in topological terms using their genus g, which, pictorially speaking, counts the number of holes. To demonstrate this, let us imagine a sphere assigned with $g = 0$ and a torus with $g = 1$. A mathematical theory of this classification of surfaces states that the integral of the Gaussian curvature over a closed surface is a quantized topological invariant, and its value is related to g. The Chern number described above can thus be regarded by analogy as the integral of a related curvature.

The exchange of the Chern number requires closing the insulating gap. This happens to the Hamiltonian (5.8) at Dirac-like points where locally cone-shaped valleys of the conduction and valence bands touch mutually. This leads to a change in the Chern number and to the related metallic boundary states protected by the still-insulating phase inside the sample [Hasan and Kane (2010); Qi and Zhang (2011)].

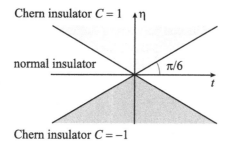

Fig. 5.10 The phase diagram for the model Hamiltonian (5.8) in a honeycomb structure; when the Chern number C equals ± 1, the topological insulator corresponds to the ground state, whereas for $C = 0$, the ground state corresponds to the ordinary insulator. *Based on [Hasan and Kane (2010)].*

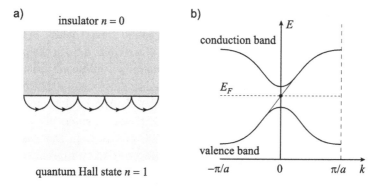

Fig. 5.11 The interface between a quantum Hall state and an insulator is assigned with a chiral edge mode; (a) the skipping cyclotron orbits are shown, which protect one direction of electron propagation along the edge (thus chiral); (b) the electronic structure of a semi-infinite strip described by the Haldane model; a single edge state connects the valence band to the conduction band. *Based on [Hasan and Kane (2010)].*

The systems that behave according to the scheme originally described by Haldane are now called Chern insulators. In these systems, the time-reversal symmetry is broken, similarly as in the IQHE, but in Chern insulators, it is broken by the presence of a net magnetic moment in the unit cell rather than by an external magnetic field, as it was the case for the IQHE. To date, Chern insulators have not been experimentally observed.

5.3.2 *Spin-Hall topological insulators*

The Hall conductivity is odd under time-reversal transform, T, and the topologically nontrivial states described above can only occur when T symmetry is broken. Nevertheless, the spin-orbit interaction allows for a different topological class of insulating band structures associated with unbroken T symmetry [Kane and Mele (2005); Moore (2009)]. The way to understand this new possible topological class is to remind the role of T symmetry for particles with spin 1/2. The time-reversion transform is represented in this case by an anti-unitary operator,

$$\Theta = i\sigma_y K, \tag{5.10}$$

where σ_y is the y component of the spin operator (y Pauli matrix) and K is the complex conjugation. For electrons with spin 1/2, $\Theta^2 = -1$ and this property lead to the Kramers theorem. According to the Kramers theorem, all eigenstates for a T-invariant Hamiltonian must be at least two-fold degenerate [2]. In the absence of spin-orbit interactions (i.e., when spin and orbital degrees of freedom are independent), Kramers degeneracy is simply the degeneracy between up and down spins. Nevertheless, in the case of coupling between spin and orbital degrees of freedom, e.g., in the presence of spin-orbit interactions, the Kramers degeneracy would lead to nontrivial consequences. A T-invariant Bloch Hamiltonian must satisfy the condition $\Theta H(\mathbf{k})\Theta^{-1} = H(-\mathbf{k})$. It is possible to classify the equivalence classes of Hamiltonians satisfying this condition that can be smoothly deformed without closing the energy gap. The Chern invariant is in this case C=0, but an additional invariant, n, can be introduced with two possible values, $n = 0$ or 1. The associated $n = 0$ or 1 values for two distinct topological classes can be noticed by referring to the bulk-boundary correspondence. In analogy to the spinless case with T symmetry broken, which is illustrated in Fig. 5.11, one can schematically depict the electronic states associated with the edge of a T-invariant 2D insulator as a function of the quasi-momentum along the edge, as in Fig. 5.12.

In Fig. 5.12, the Brillouin zone (for 1D model system) exhibits mirror-type symmetry caused by time-inversion symmetry. Thus, it is sufficient to consider only half of the Brillouin zone $[0, \pi/a]$ because T symmetry requires that the other half be a mirror image of the former. In this figure, the bulk conduction and valence bands separated by an energy gap are schematically depicted. Near the edge, there may or may not occur states bound to the edge inside the gap, depending on the details of the Hamiltonian. When such states bound to the edge inside the forbidden gap are present, the Kramers theorem forces them to be two-fold degenerate at the

---

[2]This conclusion follows from the simple observation: if $|\psi>$ is nondegenerate, then $\Theta|\psi> = c|\psi>$ and $|c|^2 = 1 \neq -1$; thus, this contradiction requires degeneracy.

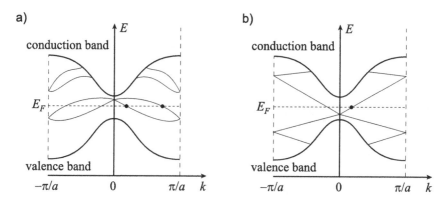

Fig. 5.12 The electron dispersion between two boundary degenerate points (due to Kramers theorem), $k_x = 0$ and π/a; in (a) the number of surface states crossing the Fermi level is even, whereas in (b) it is odd. An odd number of crossings leads to topologically protected metallic boundary states. *Based on [Hasan and Kane (2010)].*

T-invariant momenta $k_x = 0$ and π/a (the latter point is the same as $-\pi/a$ due to periodicity and simultaneously due to mirror symmetry). Away from these special points, $0, \pi/a$, a spin-orbit interaction can remove the degeneracy. In general, there exist two ways in which the states at $k_x = 0$ and π/a can be connected [Hasan and Kane (2010)]. The first possible way of connection is visualized in Fig. 5.12 (a); in this case, the points connect pairwise (in the right half of the Brillouin zone). In this particular case, the edge states can be smoothly eliminated by pushing all of the bound states out of the forbidden gap. In this case, the bands intersect the Fermi level an even number of times, between $k_x = 0$ and π/a. In contrast, in the the other possible situation, visualized in Fig. 5.12 (b), the edge states cannot be smoothly eliminated by pushing parameters. In this case, the bands intersect the Fermi level an odd number of times within the section $[0, \pi/a]$. The two distinct situations described above give rise to Z_2-valued invariant n, which can be assigned to appropriate topologically nonequivalent states in the case of T-invariant spin systems.

Interest in the spin-Hall effect has rapidly increased since 2000. The effect manifests itself by the existence of spin-polarized edge states in a sample of a semiconductor with strong spin-orbit interaction, which can be observed when an electric charge current is pushed through the sample. This effect was confirmed experimentally [Kato *et al.* (2004)]. The quantized version of a similar phenomenon was theoretically proposed for graphene, which includes topological edge modes responsible for the spin

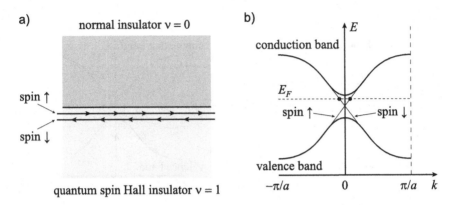

Fig. 5.13 Edge states in the quantum spin Hall insulator; the interface between a spin Hall insulator and an ordinary insulator (a); the edge state dispersion in the graphene-like model in which up and down spins propagate in opposite directions (b). *Based on [Hasan and Kane (2010)].*

current around the edges [Kane and Mele (2005)]. The emergence of the spin-carrying edge states in this model is triggered solely by the intrinsic spin-orbit interaction, and the flow of the spin current is protected by the time-reversal symmetry. Materials that are not magnetically ordered display time-reversal symmetry, and there is a relatively rich group of materials with strong spin-orbit interaction but no magnetically ordered. Therefore, the direct observation of the quantum spin-Hall (QSH) effect in real materials is very probable. In particular, graphene would be a QSH insulator, but the spin-orbit interaction is extremely weak in graphene. This makes the experimental detection of the QSH effect in this material very difficult; indeed, this feat has not yet been achieved.

In a different system, a HgTe/CdTe quantum well with particularly strong spin-orbit interaction, the QSH effect was predicted; this was also confirmed experimentally [König *et al.* (2007)] along the previously shaped theoretical scenario [Bernevig *et al.* (2006)], cf. Fig. 5.14. In this way, the first QSH insulator was discovered and the new field of topological insulators was opened. These materials are defined as insulators in the bulk but metallic along any edge or surface that is cut into the sample. Unfortunately, HgTe quantum wells remain the only two-dimensional QSH insulators that have been discovered to date. In the 3D case, opportunities for observing QSH insulators are much richer, as is listed in, e.g., Refs [Sato *et al.* (2010); Kim *et al.* (2010); Xia *et al.* (2009); Roy (2009)].

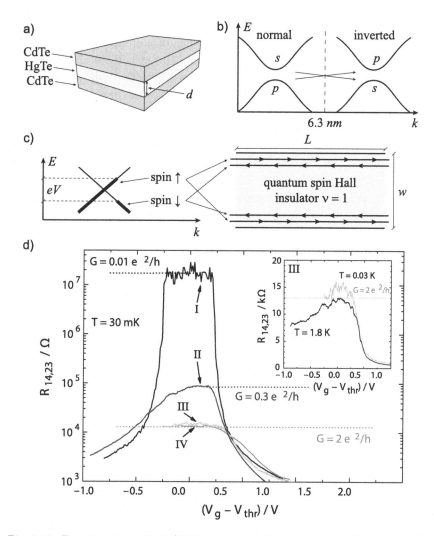

Fig. 5.14 Experiments on HgTe/CdTe quantum wells: quantum well structure (a); when varying thickness d, the 2D quantum well states cross and band inversion transition takes place (b); the inverted state is the quantum Hall insulator, which has helical edge states that have a nonequilibrium population determined by the leads (c); experimentally observed resistance as a function of a gate voltage that tunes the Fermi level through the bulk gap. Sample I, with $d < d_c$, shows insulating behavior, while samples III and IV show quantized transport associated with edge states (d). *Source: M. König, S. Wiedmann, C. Brüne, A. Roth, H. Buhmann, L. W. Molenkamp, X.-L. Qi and S.-C. Zhang, "Quantum spin Hall insulator state in HgTe quantum wells", Science 318, 766 (2007).*

Metallic states at the edge are a demonstration of the fact that the bulk insulating gap is reduced to zero at the edge of a sample. The most essential feature observed when modeling the band structures of topological insulators is the emergence of conical points where the bulk bands touch each other, reducing the gap to zero. These singular Dirac points are actually the root of the topological nontrivial properties of related models. This also reveals connections between topological insulators and the features characteristic of graphene band structure related with the presence of Dirac points in this material [Prodan (2011); Hasan and Kane (2010); Qi and Zhang (2011)].

Unfortunately, measurements reporting a clear signal coming from the metallic surface, even those performed on ultra-high-quality crystals, reveal, however, rather low surface conductivity. Therefore, this necessitates the further development and incsease of experimental resolution to unambiguously determine and confirm the expected robustness against disorder. This has to be rigorously tested, but the experimental confirmation of this property has not yet been achieved (for recent progress in this respect cf. [Noh *et al.* (2011)]).

Moreover, it should be emphasized that Chern insulators and spin Hall insulators can be understood using a single-particle approach in Bloch band Hamiltonian terms, including nontrivial topological effects familiar in the IQHE. In contrast, the FQHE is essentially a collective phenomenon, with the primary role played by interaction and with deep topological constraints, as described in the previous chapter. Thus, this effect cannot be explained using single-particle approach, as in the case of the IQHE and topological insulators. Despite the apparent strong correlations-linked background of the FQHE, one can expect, however, an influence of the ideas related in particular to spin structure and time-reversal symmetry on the understanding of FQHE ordering [Sheng *et al.* (2011); Levin and Stern (2009)], taking into account the still-growing knowledge based on spin Hall insulating states.

5.4   Topological states in optical lattices

The trapping of cold atoms in artificially arranged optical lattices opened a new avenue in experimental and related theoretical investigations of correlated multiparticle systems, including fermions, bosons and mixtures. Although trapped particles are neutral atoms as a rule, methods of including strong interaction were proposed [Sørensen *et al.* (2005); Stanescu *et al.* (2010)]. The methods of magnetic-field imitations were also developed via the rotation of unit cells in optical lattices. Special attention has been paid to studies of topological states also including the FQHE [Palmer and Jaksch (2006); Stanescu *et al.* (2010); Zhang *et al.* (2011); Wu *et al.* (2007); Sørensen *et al.* (2005); Palmer *et al.* (2008)]. The rotating optical lattice scheme used to model a magnetic field is presented in Fig. 5.15.

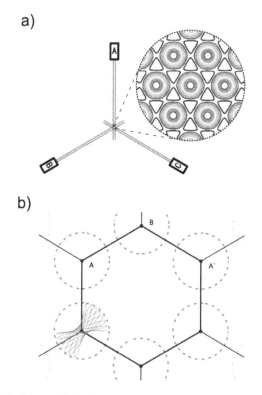

Fig. 5.15 A sketch of the method of creating an optical lattice with an artificial magnetic field. The three laser beams cross each other at 120° in the plane; phase modulators are placed in the paths of the two beams (a); an illustration of the on-site rotation [Gemelke (2007)]; the entire lattice undergoes fast oscillation and a slow precession with a frequency Ω, as schematically plotted by the red solid lines in one lattice site; after taking the time average of the fast oscillation, atoms feel each site rotate around its own center with the precession frequency Ω, which is plotted by the red dot-dashed line around each site. *Source: M. Zhang, H. Hung, C. Zhang and C. Wu, "Quantum anomalous Hall states in the p-orbital honeycomb optical lattices", Phys. Rev. A 83, 023615 (2011).*

It should be noted that experimental observations of the FQHE in optical lattices have not yet been made. Similarly, other collective states have not been arranged in practical setups, although theoretical investigations are quite advanced. Therefore, one can conjecture that the analogy of these novel experimental systems of atoms trapped in optical lattices, even with magnetic cyclotronic rotation implemented by phase-shift-induced rotation, as demonstrated in Fig. 5.15, with real multiparticle quantum systems such as 2DEG or graphene, is not complete. This may be related to the classical scenario of rotation imitating a magnetic field. When these orbits are too short for particle interchanges, as in fractional filling, multi-loop braids in 2D ought to be important, along with the scheme presented in previous chapters. In the model calculus of the quantum type, this is arranged automatically, but not in an optical lattice setup. Thus, despite many theoretical results supporting the existence of the FQHE for model Hamiltonians in optical lattices, experimental observation has remained elusive. It is likely that the development of the method of magnetic-field imitation by means of phase-induced rotation is required for the practical realization of the quantum effects of multi-looped kinetics of 2D quasiclassical wave packets.

In this regard, it is worth emphasizing that there are also recent analyses of the FQHE in model systems without Landau levels [Sun *et al.* (2011); Sheng *et al.* (2011)]. In the paper [Sun *et al.* (2011)], the authors constructed a class of model Hamiltonians on a 2D lattice that give nearly-flat bands with nontrivial topology. This property is regarded as a required prerequisite for the organization of FQHE-like states when interaction would be included next. For flat a band, the kinetic energy is frozen, which allows the interaction to dominate and create strongly correlated states of the FQHE form. This has been theoretically verified in the paper [Sheng *et al.* (2011)] for a checkerboard lattice including nearest and next-nearest interaction terms in the model Hamiltonian of the following form,

$$H = -H_0 + U \sum_{<i,j>} n_i n_j + V \sum_{\ll i,j \gg} n_i n_j, \qquad (5.11)$$

where H_0 is a two-band checkerboard lattice model with nonzero Chern number implemented by complex hopping factors of the type,

$$H_0 = -t \sum_{<i,j>} e^{i\psi_{ij}} (c_i^+ c_j + h.c.) + H_1, \qquad (5.12)$$

and H_1 describes ordinary real-number-assigned hopping between next and next-next nearest neighbors [Sun *et al.* (2011)]. Some attributes characteristic of the FQHE have been indicated using this model for fractional fillings

1/3 and 1/5 (in the latter case, repulsion V was required above a certain threshold value, unless U is strongly enhanced) [Sheng *et al.* (2011)]. This indication supports the idea of the multi-looped structure of quasiclassical wave packets, which in this way can reach equidistantly separated 2D particles due to the repulsion interaction in an almost flat band with suppressed kinetic energy, quite similar to the situation described above for a 2DEG system in a magnetic field. Although any links can not be drawn here to the CF model with auxiliary magnetic field flux quanta attached to hypothetical composite particles, the multi-looped requirements for enhancing orbits still hold. Because the number of loops can only be an integer, this explains the fractional structure of fillings exactly in the same manner as in the case of a 2DEG under a strong magnetic field. The role of cyclotron orbit quantization is substituted here by orbit quantization due to Chern number (Berry phase) invariant conservation. Although the theoretical models of the FQHE or similar topological states without Landau levels, also for almost flat model bands, can be referred to optical lattice experiments, no experimental confirmation has yet been achieved. One could suspect that progress would require the further development of artificial modeling (by rotation due to phase shifts in laser beams creating the lattice) toward multi-looped structures of effective trajectories instead of the single-looped structures that are currently attainable.

Chapter 6

Summary

A not-simply connected (multiply connected) structure of configuration spaces for multiparticle systems is expressed by braid groups, i.e., by the first homotopy groups π_1 of these spaces, which are non-trivial groups in such cases. Braid groups are extremely complicated in the case of 2D (and locally 2D) systems. A braid presentation indicates that trajectories in a configuration space of a multiparticle system cannot be continuously deformed or transformed from one into another if they belong to different classes of braid group homotopy. For that reason, it is not possible to integrate over trajectories (in path integrals) using a single measure for whole trajectory space, and the required continuity conditions allow for the definition of measures in the segments of trajectory space only for mutually homotopic trajectories, i.e., separately for every class of homotopy. This leads to dividing the field of Feynman integrals over trajectories according to the structure of a full braid group. Therefore, the need to determine the one-dimensional unitary representation for this group arises to define weights, with which individual mutually nonhomotopic classes of multiparticle trajectories would be included in functional integration. Each of these representations provides ground for characterizing a separate type of quantum particle with different statistics. In the case of R^3, and also of manifolds with more dimensions, the full braid group is a permutation group, which has only two different one-dimensional unitary representations, leading to bosons and fermions. In the case of a R^2 manifold (and also compact locally 2D manifolds), the far richer unitary representations of the braid group allow for the existence of anyons apart from bosons and fermions.

In this treatise, we argue that the sum over nonhomotopic trajectory classes in functional integrals can only include those elements of the full braid group that describe possible trajectories. We demonstrate that in the presence of a strong magnetic field perpendicular to a system of 2D charged particles, not all trajectories defined for a 2D system without the field would be available. In the case when the cyclotron radius is too short compared to the distance between particles, which is sustained by Coulomb repulsion at the level determined by the system's density, particle exchanges along cyclotron trajectories are impossible (as is 2D long-range helicoidal motion is impossible), and the corresponding elements of the braid group should be excluded from the corresponding path integrals. Thus, the remaining braid subgroup of trajectory classes, referred to here as the cyclotron braid group, describes the domain of functional summation and, through its unitary representations, indicates the possible quantum particles in two-dimensional charged systems in the presence of a strong magnetic field. The appropriate one-dimensional unitary representations of the cyclotron braid subgroups allow for the identification of composite fermions (more generally, composite anyons), particles previously modeled using artificial constructions with auxiliary local flux tubes or vortices and present in Hall systems with Laughlin correlations.

These particles, observed experimentally in the FQHE, could not be distinguished from ordinary fermions by representations of the full braid group (because of the 2π periodicity of the phase factor $e^{i\alpha}$), and only after introducing the braid structure characteristic to the presence of a strong magnetic field, in the form of the braid cyclotron subgroup, one can properly describe composite fermions in clear association with the required Laughlin correlations. In this way, we are able to identify the topological conditions for the exotic physics of 2D Hall systems without referring to auxiliary effective constructions. It has been observed that in the case of charged multiparticle 2D systems in the presence of a strong magnetic field, the multi-looped cyclotron structure of braid trajectories meets the expected requirements for particle exchanges (when single-looped cyclotron trajectories are too short in comparison to interparticle separation). This is related to the fact that in planar systems, additional loops cannot lead to an increase in area (in contrast to 3D systems); thus, they must share the same external field flux, which leads to an increase in the cyclotron loop size to a degree of interparticle separation and thereby enables particle exchanges necessary to establish the related statistics (via the appropriate definition of the braid group, in this case called the cyclotron braid subgroup).

At the same time, the details and the character of model auxiliary objects in previous formulations referring to composite fermions and Laughlin correlations have been explained, particularly the fictitious character of magnetic flux tubes attached to particles in the model of composite fermions, which has been introduced to increase the cyclotron radius in reduced field by these flux tubes oppositely oriented with respect to the external field. Attention has been drawn to the fact that in this model approach, the arbitrary assumption regarding the total number of flux tube quanta attached to particles for Landau level fillings also apart $\nu = \frac{1}{p}$, p odd, leads to an unclear possibility of turning the effective resultant field of the external field and of the average field of attached flux tubes, opposite to the direction of the external field, which was assumed by Jain to restore the hierarchy of fillings for the FQHE. The cyclotron formulation eliminates these doubts from the description of Laughlin correlations with the use of composite fermions and admits an improved hierarchy of fillings for the FQHE by mapping onto the integer Hall effect without unexplained assumptions.

Another possible source of discrepancy with respect to a traditional formulation of composite fermion theory has also been indicated when during numerical diagonalizations of interaction (widely applied to Hall system analyse), the domain for minimization has not been limited to Hilbert's subspace of anti-symmetric functions with the required Laughlin symmetry adjusted to composite fermions as rightful quantum particles separated in quantum-statistics terms from ordinary fermions. Composite fermions were misleadingly treated in traditional formulation as ordinary fermions only dressed, due to Coulomb interaction, with local flux tubes in some analogy to quasiparticles familiar in solids. This may lead to a significant inaccuracy as a result of minimization over the excessively large domain of all antisymmetric functions corresponding to ordinary fermions, instead of much smaller domain corresponding to composite fermions with selected Laughlin symmetry determined by appropriate phase shift $p\pi$ and not π (which makes a difference in 2D, where the braid group is not a permutation group).

The formalism of cyclotron braid subgroups suggested in this treatise explains in a consistent way the reason for exotic and complex Laughlin correlations in charged 2D multiparticle systems in a strong magnetic field. It also enables us to explain and verify the widely used descriptions of systems with the use of composite fermions. In particular, recent experimental observations of the fractional quantum Hall effect in graphene, especially in competition with the insulating state, strongly support the role of the mobility of carriers and not only of interaction. This agrees with the cyclotron braid picture, which also emphasizes the role of high mobility in the creation of fractional quantum Hall collective states. For graphene, such states appear only in samples with strongly enhanced mobility (as in so-called suspended graphene), and the triggering role of the mobility enhancement via sample annealing was directly demonstrated. The explanation of the significance of the mobility in the formation of the FQHE, apart from that of particle interaction, is clearly beyond the traditional single-particle picture of composite fermions as particles dressed with hypothetical flux tubes due to interaction. The relevant explanation requires the involvement of nonlocal topology-type argumentation in line with the ideas linked to quasiclassical wave-packet trajectories, specifically in 2D under a strong magnetic field at fractional LLs fillings, which would be expressed in terms of also quasiclassical properties such as the carrier mobility or flux quantization. Progress in terms of the fundamental physics and topological conditions of quantization of 2D charged systems in a magnetic field, obtained as a result of considering cyclotron braid subgroups, suggests the immense usefulness of homotopy methods in the analysis of multiparticle quantum systems with various geometries. The related link between the quantum properties of peculiar geometry-confined systems (such as 2D charged systems in the presence of a strong magnetic field) and their topology, expressed with the help of Feynman path integral methods as demonstrated above, has a general significance in the recognition of quantum effects caused by the topology, which is essentially classical. The central role play here the braid groups, which are well recognized in mathematic side within rapidly grown up in the past few decades the homotopy approach in algebraic topology. Topology terms have become increasingly important in the analyses of complicated, strongly correlated systems. A good example of this is the outburst and then rapid development of interest for so-called topological insulators, perceived as a new state of matter. High expectations for new solutions, and of possible practical applications, are linked also to decoherence-free topology quantum information processing schemes employing nonabelian anyons described by multidimensional unitary representations of braid groups in 2D. The basic relations between physics and information processing allow us to expect an even wider use of braid groups (full, cyclotron and pure braid groups alike) in both of these fields.

Chapter 7

Comments and supplements

7.1 The wave function for a completely filled lowest Landau level

The Landau quantization of the dynamics of a charged particle in a magnetic field in a plane perpendicular to the field's direction is an exact result with numerous significant consequences, including the physics of multiparticle systems, such as electrons in metals or semiconductors, or in 2D Hall systems. If we consider a single charged particle (whose statistics is of no importance and cannot be verified via particle interchanges in the case of a single particle), then, at a constant magnetic field presence, it corresponds to a Hamiltonian,

$$H = \frac{(-i\hbar\nabla - \frac{e}{c}\mathbf{A}(\mathbf{r}))^2}{2m} - 2\mu_B \mathbf{B} \cdot \mathbf{S}, \qquad (7.1)$$

where e stands for the particle's charge, $\mathbf{A}$ is the vector potential of a magnetic field, $\mathbf{B} = \nabla \times \mathbf{A}$ in the Pauli term [1], $-2\mu_B\mathbf{B} \cdot \mathbf{S}$, where μ_B Bohr magneton, $\mu_B = \frac{e\hbar}{2mc}$, $\mathbf{S} = \frac{1}{2}(\sigma_x, \sigma_y, \sigma_z)$ spin operator (we have assumed a spin of $\frac{1}{2}$, as for an electron) expressed by Pauli matrices [2]. If we choose the gauge in the following form (Landau gauge):

$$\mathbf{A} = (0, Bx, 0),$$

then the equation for stationary states attains the following form,

$$-\frac{\hbar^2}{2m}\frac{\partial^2\psi}{\partial x^2} + \frac{1}{2m}\left(-i\hbar\frac{\partial}{\partial y} - \frac{e}{c}Bx\right)^2\psi - \frac{\hbar^2}{2m}\frac{\partial^2\psi}{\partial z^2} = \varepsilon\psi, \qquad (7.2)$$

---

[1] The Pauli term ioften omitted because, in the condensed phase systems, it is multiplied by the gyromagnetic factor, usually a small one (e.g., in GaAs, the Zeeman splitting caused by the Pauli term is within the order of magnitude of 0.1 meV/T).

[2] The Pauli term in the Hamiltonian only causes a homogeneous shift of the entire spectrum by $\mp\mu_B B$ for spins along and opposite to the field (the direction z is chosen along the field $\mathbf{B}$) and therefore plays no major role in the Landau quantization.

and the eigenfunction should be chosen in the form,

$$\psi(x, y, z) = e^{ip_y y/\hbar} e^{ip_z z/\hbar} \varphi(x). \tag{7.3}$$

Therefore, the function $\varphi(x)$ satisfies the equation,

$$-\frac{\hbar^2}{2m} \frac{d^2 \varphi(x)}{dx^2} + \frac{e^2 B^2}{2mc^2} \left(x - \frac{p_y c}{eB} \right)^2 \varphi(x) = \left(\varepsilon - \frac{p_z^2}{2m} \right) \varphi(x). \tag{7.4}$$

This is an equation for quantum one-dimensional harmonic oscillator [Landau and Lifshitz (1972)],

$$-\frac{\hbar^2}{2m} \frac{d^2 \varphi}{dx^2} + \frac{kx^2}{2} \varphi = E\varphi,$$

with the center shifted by $\frac{cp_y}{eB}$ and the frequency $\omega = \sqrt{\frac{k}{m}} = \frac{Be}{mc}$ and $E = \varepsilon - \frac{p_z^2}{2m}$. Thus, we obtain the spectrum:

$$\varepsilon = \frac{p_z^2}{2m} + \mu_B B(2n + 1) \pm \mu_B B, \tag{7.5}$$

where $\pm \mu_B B$ corresponds to the energy shift for two spin orientations (due to the Pauli term in the Hamiltonian). If we omit this term as well as confine ourselves to a 2D case, we will receive an oscillator spectrum,

$$\varepsilon = \mu_B B(2n + 1)$$

and eigenfunctions in the form of harmonic oscillator states (with Hermite polynomials H_n) with the center shifted along the direction x by the value $x_0 = \frac{ep_y}{eB}$:

$$\varphi_n(x) = \frac{1}{\pi^{1/4} a^{1/2} \sqrt{2^n n!}} e^{(x - x_0)^2 / 2a^2} H_n \left(\frac{x - x_0}{aB} \right). \tag{7.6}$$

The shift of the center of the oscillator is the reason of the degeneracy of all levels to the same degree each, $n = 0, 1, 2, 3....$ This degeneracy can be easily calculated, because, $0 < \frac{cp_y}{eB} < L_x$ and $dn_y = dp_y \frac{L_y}{\hbar}$, which gives us the degeneracy of each level, $\frac{L_1 L_2 B}{hc/e} = \frac{BS}{hc/e}$ (i.e., the size of the flux of the field BS expressed by flux quanta $\frac{hc}{e}$).

For other gauges, corresponding e.g., to a cylindrical symmetry (in the cylindrical coordinates),

$$A_\phi = B\rho/2, \; A_\rho = A_z = 0,$$

the Schrödinger equation takes the form,

$$\begin{aligned} -\frac{\hbar^2}{2m} \left[\frac{1}{\rho} \frac{\partial}{\partial \rho} \left(\rho \frac{\partial \psi}{\partial \rho} \right) + \frac{\partial^2 \psi}{\partial z^2} + \frac{1}{\rho^2} \frac{\partial^2 \psi}{\partial \phi^2} \right] \\ -\frac{i\hbar\omega}{2} \frac{\partial \psi}{\partial \phi} + \frac{m\omega^2}{8} \rho^2 \pi = E\psi, \end{aligned} \tag{7.7}$$

where $\omega = \frac{|e|B}{mc}$. The solution has the form,

$$\psi = \frac{1}{\sqrt{2\pi}}e^{iM\phi}e^{ip_z z/\hbar}R(\rho),$$

where the radial part satisfies the equation,

$$\zeta R'' + R' + \left(-\frac{\zeta}{4} + \beta - \frac{M^2}{4\zeta}\right)R = 0, \qquad (7.8)$$

where, $\beta = \frac{1}{\hbar\omega}\left(E - \frac{p_z^2}{2m}\right) - \frac{M}{2}$, $\zeta = \frac{m\omega}{2\hbar}\rho^2$. For $\zeta \to \infty$, asymptotic R has the form $e^{-\zeta/2}$, and for $\zeta \to 0$, the form $\zeta^{|M|/2}$. Therefore, $R(\zeta) = e^{-\zeta/2}\zeta^{|M|/2}w(\zeta)$, and then w satisfies the equation for the confluent hypergeometric function [Landau and Lifshitz (1972); Eliutin and Krivchenkov (1976)],

$$w = F\left\{-\left(\beta - \frac{|M|+1}{2}\right), |M|+1, \zeta\right\}, \qquad (7.9)$$

where $\beta - (|M|+1)/2$ must be a non-negative integer n_ρ and the eigenvalues are given by the formula,

$$E = \hbar\omega\left(n_\rho + \frac{|M| + M + 1}{2}\right) + \frac{p_z^2}{2m}. \qquad (7.10)$$

These are the same eigenvalues as in the Landau gauge above, but are expressed with other quantum numbers (M is the z-th element of the momentum). The degeneracy is also the same as before, but here, it is expressed by an independence of E from the negative M. The wave functions have the following form,

$$R_{n_\rho M}(\rho) = \frac{1}{a^{1+|M|}}\left[\frac{(|M|+n_\rho)!}{2^{|M|}n_\rho!|M|!}\right]^{1/2}e^{-\rho^2/(4a^2)}\rho^{|M|} \qquad (7.11)$$
$$\times F(-n_\rho, |M|+1, \rho^2/(2a^2)),$$

which is completely different than in the case of the Landau gauge. The relationship between the forms of wave functions and the choice of the gauge of the magnetic fields potential is a manifestation not only of the unphysical character of the gauge, but also of the wave function adjustment to the gauge symmetry and to the appropriate boundary conditions, e.g., of type of a rectangula plaquette for the Landau gauge and of a circle (in case of 2D) for cylindrical symmetry. Therefore, the probability density for finding a particle (wave function module) is also not related to the classical cyclotron trajectories. Wave functions can even have no cylindrical symmetry, as in the case of the Landau gauge. This is related to the degeneration of Landau levels and the non-commutation of coordinate operators

x and y of the position of the center of cyclotron orbits (which cannot both be determined simultaneously) [Landau and Lifshitz (1972); Eliutin and Krivchenkov (1976)]. Quantum dynamics, however, exhibits the periodic cyclotron character, which can be observed in the time evolution in the Heisenberg representation of position operators, corresponding to the classical cyclotron trajectory (which means that any wave packet will undergo periodic cyclotron dynamics on a plane perpendicular to the magnetic field) [Eliutin and Krivchenkov (1976)]. Energy eigenvalues (although they can be expressed through various quantum numbers) are independent of the particular gauge choice, similar to the degeneration of energy levels. These levels are called Landau levels.

One can now consider a gas of non-interacting fermions filling degenerated Landau levels. The case of full filling of the lowest Landau level is particularly interesting, when the number of N particles is equal to $\frac{BS}{hc/e}$, i.e., the levels degeneracy. The appropriate wave function takes the form of the Slater determinant of various single-particle states within the first Landau level, which is antisymmetric due to the indistinguishability of fermions. In cylindrical symmetry, assuming that $n_\rho = 0$, the wave functions (7.11) have the following form: $\sim z^{|M|} e^{-|z|^2/2a^2}$, for $-M = 0, 1, 2, 3, ..., N - 1$ (which corresponds to the degeneration for the negative M), $z = \rho e^{i\phi} = x + iy$ [3]. Due to antisymmetry, we get the Slater determinant, i.e., the wave function

---

[3]It is customary to introduce the following notation,

$$\partial = \tfrac{\partial}{\partial z} = \tfrac{1}{2}\left(\nabla_x - i\nabla_y\right), \quad z = x + iy,$$
$$\bar{\partial} = \tfrac{\partial}{\partial \bar{z}} = \tfrac{1}{2}\left(\nabla_x + i\nabla_y\right), \quad \bar{z} = x - iy,$$

with properties, $\partial z = \bar{\partial}\bar{z} = 1$ and $\partial\bar{z} = \bar{\partial}z = 0$. Then, for the gauge $\mathbf{A} = \frac{B}{2}(-y, x)$, the annihilation and creation operators for harmonic oscillator have the form,

$$a = \sqrt{\tfrac{c}{2ehB}}\left(p_x - \tfrac{e}{c}A_x + ip_y - i\tfrac{e}{c}A_y\right) = -i\sqrt{\tfrac{\hbar c}{2eB}}\left(2\bar{\partial} + \tfrac{eB}{2\hbar c}z\right),$$
$$a^+ = \sqrt{\tfrac{c}{2ehB}}\left(p_x - \tfrac{e}{c}A_x - ip_y + i\tfrac{e}{c}A_y\right) = -i\sqrt{\tfrac{\hbar c}{2eB}}\left(2\partial - \tfrac{eB}{2\hbar c}\bar{z}\right),$$

and the condition for the ground state can be written as follows,

$$a\Psi(z, \bar{z}) = -i\sqrt{\tfrac{\hbar c}{2eB}}\left(2\bar{\partial} + \tfrac{eB}{2\hbar c}z\right)\Psi(z, \bar{z}) = 0,$$

which gives the solution,

$$\Psi(z, \bar{z}) = f(z)e^{-eB\bar{z}z/4\hbar c},$$

where $f(z)$ is an arbitrary analytical function of z. One can chose the independent functions $f(z)$ to be just z^n.

for N non-interacting particles in the form,

$$\Psi(z_1,...,z_N) = const.e^{-\sum_i |z_i|^2/2a^2} \begin{vmatrix} 1 & z_1 & z_1^2 & \cdots & z_1^{N-1} \\ 1 & z_2 & z_2^2 & \cdots & z_2^{N-1} \\ \cdots & \cdots & \cdots & \cdots & \cdots \\ 1 & z_N & z_N^2 & \cdots & z_N^{N-1} \end{vmatrix}. \qquad (7.12)$$

The above determinant in the Slater function is a well-known Vandermonde determinant and the whole function can be written in the form,

$$\Psi(z_1,...,z_N) = const.e^{-\sum_i |z_i|^2/2a^2} \prod_{i,j=1,\, i>j}^{N} (z_i - z_j), \qquad (7.13)$$

where the last product is the Vandermonde polynomial. Replacing the Vandermonde polynomial with an antisymmetric Jastrow polynomial, $\prod_{i,j=1,\, i>j}^{N}(z_i - z_j)^p$ (p odd integer) changes the Slater function into the Laughlin function for FQHE at fractional fillings $\nu = \frac{1}{p}$ of LLL.

7.2 Paired Pfaffian states

The determinant for the anti-symmetric matrix $2n \times 2n$ can be expressed (with accuracy up to the sign) in the form of a square polynomial of n-th degree, built from the elements of this matrix. This polynomial is called the Pfaffian of the matrix (antisymmetric, of even order). For instance,

$$A = \begin{bmatrix} 0 & a \\ -a & 0 \end{bmatrix},$$

$$Pf(A) = \pm\sqrt{det(A)} = \pm\sqrt{a^2} = a,$$

(the sign can be determined via the definition (7.14) because using the relation $(Pf(A))^2 = det(A)$, it is possible to determine the Pfaffian with accuracy only up to the sign);

$$B = \begin{bmatrix} 0 & a & b & c \\ -a & 0 & d & e \\ -b & -d & 0 & f \\ -c & -e & -f & 0 \end{bmatrix},$$

$$Pf(B) = \pm\sqrt{det(B)} = af - be + dc.$$

It is easy to show the simple properties of the Pfaffian,

$$(Pf(A))^2 = det(A),$$

$$Pf(BAB^T) = det(B)Pf(A),$$

$$Pf(aA) = a^n Pf(A),$$

$$Pf(A^T) = (-1)^n Pf(A),$$

where A, B are antisymmetric matrices of order $2n$, where a is an arbitrary number .

The following representation of the Pfaffian of the matrix is convenient, $A = \{a_{ij}\}$ of the $2n$ order, $a_{ij} = -a_{ji}$,

$$Pf(A) = \frac{1}{2^n n!} \sum_{\sigma \in S_{2n}} sign(\sigma) \prod_{i=1}^{n} a_{\sigma(2i-1)\sigma(2i)}, \tag{7.14}$$

where S_{2n} is the permutation group, and $sign(\sigma)$ means the permutation sign (parity signature). For antisymmetric matrices, for which the Pfaffian is determined, we can introduce the pairing of indices $1, ..., 2n$, i.e.,

$$\alpha = \{(i_1, j_1), (i_2, j_2), ..., (i_n, j_n)\}, \tag{7.15}$$

where, $i_k < j_k$, and $i_1 < i_2 < ... < i_n$. Such an arbitrary pairing corresponds to the permutation:

$$\pi = \left\{ \begin{matrix} 1 & 2 & 3 & 4 & ... & 2n \\ i_1 & j_1 & i_2 & j_2 & ... & j_n \end{matrix} \right\}. \tag{7.16}$$

Using the above pairings, one can write,

$$Pf(A) = \sum_{\alpha} sign(\pi) a_{i_1 j_1} a_{i_2 j_2} ... a_{i_n j_n}, \tag{7.17}$$

where the summation is carried out over all possible pairs that meet the requirements of the indices ordering.

For instance, for the matrix 4×4, we have,

$$Pf(A) = a_{12}a_{34} - a_{13}a_{24} + a_{14}a_{23},$$

which is consistent with the root of the determinant of the matrix A.

Pfaffians turn out useful for modeling wave functions of paired $2n$ fermions in the BCS state in a position representation (in the 2D case), in the form of the function,

$$Pf\left(\frac{1}{z_i - z_j}\right). \tag{7.18}$$

Generally speaking, a BCS function in position space (for $2n$ particles in 2D) has the form [Greiter *et al.* (1992)]

$$\Psi_{BCS}(z_1, ..., z_{2n}) = Antisymmetrization \left\{ \prod_{i,even}^{2n} \psi(z_{i-1} - z_i) \right\}, \quad (7.19)$$

where *Antisymmetrization* refers to all possible pairings of $2n$ particles (there is of number $(2n-1)!!$ of these pairings), which is precisely the same as in the Pfaffian definition [Dyson F. quoted in Schrieffer J. R. (1983)]. A function of a pair of fermions $\psi(z_i - z_j)$ must be odd and should be a monotonically decreasing function of the distance between the particles, if it is to represent a pairing. The easiest such function (and the one corresponding to the standard BCS representation in the momentum space with a constant pairing potential)[4] is,

$$\psi(z_i - z_j) = \frac{1}{z_i - z_j}. \quad (7.20)$$

The wave function in the Pfaffian form (7.18) has a singularity for $z_i = z_j$, but it usually occurs as a factor together with the Jastrow polynomial, which eliminates the singularity in the entire modeled function describing the pairing, as in the case below (for LLL filling $\nu = \frac{1}{2}$),

$$\Psi(z_1, ..., z_{2n}) = Pf\left(\frac{1}{z_i - z_j}\right) \prod_{i<j} (z_i - z_j)^2 e^{-\sum_j |z_j|^2/(2a^2)}. \quad (7.21)$$

It is worth noting that he Pfaffian introduces a $-\pi$ phase shift to the above function when the particle exchange, which, together with 2π from the Jastrow polynomial, gives us the total phase shift of π (as for ordinary fermions).

7.2.1 *Fermi sea instability toward the creation of Cooper pairs in the presence of particle attraction*

The core of the superconductivity BCS theory is located in the occurrence of an instability of a two-particle Green function of a *normal* Fermi liquid in the particle-particle channel for interaction [Abrikosov *et al.* (1975)]. This singular coherent scattering channel can be observed in the Bethe-Salpeter type equation for the vertex functions, as sketched in Fig. 7.1.

---

[4]For the BCS Hamiltonian, $H = \sum_i \frac{p_i^2}{2m} - \frac{\pi}{m} \sum_{i \neq j} \delta(\mathbf{r}_i - \mathbf{r}_j)$, the paired BCS function (in 2D) has the form of $Pf\left(\frac{1}{z_i - z_j}\right)$ (which is associated with the relationship $\nabla^2 \frac{1}{z} = -2\pi\delta(z)/z$) [Greiter *et al.* (1992)].

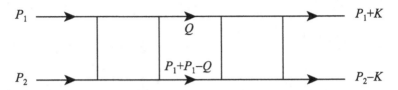

Fig. 7.1 Divergent particle-particle channel of scattering for attracting fermions corresponding to the Cooper pairing instability of the ground state of a non-interacting system. The lines represent Green functions (with Matsubara frequency and momentum argument, $P = (i\omega_\nu, \mathbf{p})$), the squares represent a bare vortices (an interaction) and the arising pole indicates instability for $\mathbf{p}_1 \simeq -\mathbf{p}_2$, at a temperature $T < T_c$ and for an arbitrary small particle attraction, cf. [Abrikosov *et al.* (1975)].

A full vertex is expressed by an irreducible vertex function with respect to two parallel-directed Green functions (particle-particle channel for scattering; in other words, one cannot cross this irreducible vertex function by that pair of parallel-oriented Green functions), or by other irreducible vertex function with respect to two anti-parallel-directed Green functions. Separate summations of an infinite graph ladder with a bare interaction corresponding to those irreducible vertices do not involve all possible scatterings, but the simultaneous summation of both infinite graph ladders is impossible. Therefore, the singularities of both channels are analyzed independently. On the one hand, the particle-hole channel corresponds to the theory of a normal Fermi liquid, it describes the collective excitations of zero-sound type and Landau damping behavior (in terms of Landau amplitudes expressed by an appropriate limit of the vertex function) [Abrikosov *et al.* (1975)]. On the other hand, the direct analysis of a particle-particle channel singularity indicates an instability of the normal Fermi liquid ground state against the pairing, which is connected with coherent propagation of the particle pair with total momentum close to zero (i.e., the particle pair taken from opposite sides of the Fermi surface), and therefore indicates the creation of Cooper pairs. From a simple analysis of the kernel of the related integral equation for a particle-particle channel (parallel-oriented Green functions), the *purely imaginary pole* of the vertex function occurs (and thus of two-particle Green function). This pole indicates the instability of the normal Fermi liquid ground state, because it occurs in the upper half-plane (after analytic continuation connected with a standard transition between Matsubara-type functions to retarded/advanced Green functions), however, under the following conditions: 1) an attractive nature of interaction (sign minus for the bare vertex function), 2) a small overall momentum of both coherent scattering particles, 3) a sufficiently low temperature (conditions 2 and 3 lead to the possibility of defining the critical value of the Cooper pairs momentum and the critical temperature for the instability of the normal state corresponding to the creation of those pairs) [Abrikosov *et al.* (1975)] chapter VII $33. In needs to emphasized that the described in this manner instability of the normal Fermi sea in case of an (arbitrarily small) attraction indicates the coherent formation of Cooper pairs, thus, it is linked with the tendency to reconstruct the ground state of the normal Fermi liquid toward the BCS state (identified, however, from the side of the *normal* state). A complete description of the superconducting state (or superfluid in the case of chargeless fermions), which requires introducing anomalous Green functions and anomalous vertices and thus exceeds beyond the frames of normal Fermi liquid theory, was formulated by Larkin and Migdal for type-s of pairing and by Czerwonko for type-p of pairing, as is described with the relevant bibliography in e.g., [Abrikosov *et al.* (1975); Jacak (1988)].

7.3 Basic definitions in group theory

Subgroup   The set $G' \subseteq G$ is called the subgroup of the G group when

(1) $e \in G'$,

(2) $\underset{x,y \in G'}{\forall} \ x \bullet y \in G'$,

(3) if $x \in G'$ to $x^{-1} \in G'$,

i.e., the pair $\{G', \bullet\}$ is a group.

Group order   The order of the group G is the number of elements in that group, denoted as n. If n is finite $(n < \infty)$, then the group has a finite order; if $n = \infty$, then the group has an infinite order.

Cyclic groups   When all elements of the group $G = \underbrace{\{e, a, b, \ldots\}}_{n}$ can be written in the form of $G = \{e, a, a^2, a^3, \ldots, a^{n-1}\}$, where $a^k = \underbrace{a \bullet \ldots \bullet a}_{k}$, $k \in \{0, \ldots, n\}$ ("$\bullet$" is a group action) and $a^k = e \Leftrightarrow k = 0 \lor k = n$, then such G group is called cyclical and has n order.

Permutation groups   The group $S(G)$ of all bijection mappings of a n-element set G create a symmetric group S_n. The symmetric group S_n is a permutation group of n elements. The order of S_n is $n!$.

Group generators   Let X be any subset of the G group. By (X) we indicate a cross section of all subgroups within the G group involving X. (X) is a subgroup of the G group and is called the subgroup generated by X. Elements of the X set are called group generators (X). The subgroup (X) consists of elements in the form:

$$x_1^{\alpha_1} \cdot x_2^{\alpha_2} \cdot \ldots \cdot x_n^{\alpha_n},$$

where $x_i \in X.$, $\alpha_i = \pm 1$.

If the group G is generated by a finite set of generators X, then it cannot be uncountable.

Layers   Let G be a group and G' be its subgroup. *The left-hand layer* of the G' subgroup in the G group stands for a set $xG' = \{x \bullet g, g \in G'\}$, where $x \in G$. Each element of the layer xG' is its representative. *The right-hand layer* of the G' subgroup in the G group stands for a set $G'x = \{g \bullet x, g \in G'\}$, where $x \in G$. Each element of the layer $G'x$ is its representative.

Group homomorphism   Homomorphism $f : \{G, \bullet\} \to \{G', \times\}$ (denoted as $f : G \to G'$ or f) is a mapping which sustains group action, i.e.,

$$\underset{x,y \in G}{\forall} f (x \bullet y) = f (x) \times f (y).$$

Group homomorphism kernel   The kernel of homomorphism $f : G \to G'$ is the set $Ker f = \{x \in G, f(x) = e'\}$, where e' is the neutral element of G'.

Suppose $H = Ker f$

- **Lemma 1.** The kernel of H homomorphism $f : G \to G'$ is a subgroup of the G group.
- **Lemma 2.** $\underset{x \in G}{\forall}$ $xH = Hx$, where $xH = \{x \bullet h, h \in H\}$ and $Hx = \{h \bullet x, h \in H\}$.
- **Lemma 3.** A pair $\{H, \bullet\}$ is a group.
- **Lemma 4.** The map $f' : G' \to G$ is a homomorphism.
- **Lemma 5.** The subgroup $H \subset Ker f'$, or even $H = Ker f'$.

Quotient group denoted as G/H is a group of layers G' of the normal divisor H.

Normal divisor (normal subgroup) The normal divisor is any subgroup G' of the G group that meets the following requirement:

$$\bigvee_{x \in G} xG'x^{-1} = G'.$$

Every divisor is a kernel of some homomorphism.

Group representation The representation of the G group is the homomorphic mapping of this group into a set of finite-dimensional square matrices [5]. The group, whose elements are matrices $D(x)$, $x \in G$, which is a homomorphic image of the G group, is its representation. There are always trivial representations (of any dimension) of the G group, which are identity matrices.

If A is a representation of the G group, then the determinants $\det D(x)$, $D(x) \in A$, $x \in G$ create a single-dimensional representation of the G group. If a homomorphism exists, $f : G \to G'$, and A is a representation of the group G', then A is a representation of the group G. Suppose $D(x)$ and $D'(x)$, where $x \in G$ will be representations of the G group. If $D'(x) = S^{-1}D(x)S$, where S is any nonsingular matrix (thus describing exchange of the basis in V space, for orthonormal bases the matrix S is unitary one), then $D(x)$ and $D'(x)$ are called equivalent or similar representations. Similar matrices represent the same linear transformation but after a change of the basis (for the domain and range simultaneously).

Irreducible representations of the group An irreducible representation of a group is a group representation that has no nontrivial invariant subspaces. A subspace of the space V that is invariant under the group action makes opportunity to define a subrepresentation. If V has exactly two such subspaces, namely the zero-dimensional subspace and V itself, then the representation is said to be irreducible; if it has, however, a proper subspace of nonzero dimension invariant under the group action, the representation is said to be reducible.

The irreducible representation is a representation that cannot be reduced to simpler block form with the use of a similarity transformation [6].

---

[5] It is a homomorphism of the group into space of linear auto-transformations of n-dimensional vector space V, $\phi; G \to GL(V)$, thus elements of G group are represented by $n \times n$ matrices in a particular basis of n-dimensional space V.

[6] Representation of groups are considered as linear auto-transformations of multi-dimensional linear spaces, thus via changing bases in these spaces, one can transform matrices of group representation to the diagonal form or, more generally, to the Jordan canonical block form.

Representations of finite groups over the complex numbers can be decomposed into a direct sum of irreducible subrepresentations (Maschke theorem [Lang (2002); Serre (1977)]),

$$D\left(x\right) = a_1 D_1\left(x\right) \oplus a_2 D_2\left(x\right) \oplus \ldots \oplus a_n D_n\left(x\right) = \overset{n}{\underset{i=1}{\oplus}} a_i D_i\left(x\right),$$

where $D_i\left(x\right)$ means the irreducible representation and a_i is the number of subspaces that transform according to the same irreducible representation (this means that an irreducible representation $D_i\left(x\right)$ is included in the reducible representation a_i times). The representation $D\left(x\right)$ can be thus presented in a quasidiagonal form,

$$D\left(x\right) = \begin{bmatrix} \begin{bmatrix} \\ \end{bmatrix} & 0\,0\,0\ 0\,0 \\ & 0\,0\,0\ 0\,0 \\ 0\,0 & \begin{bmatrix} \\ \end{bmatrix} & 0\,0 \\ 0\,0 & & 0\,0 \\ 0\,0 & \begin{bmatrix} \\ \end{bmatrix} & 0\,0 \\ 0\,0\ 0\,0 & & \\ 0\,0\ 0\,0 & & \ddots \end{bmatrix}$$

$$= a_1 \begin{bmatrix} \\ \end{bmatrix} \oplus a_2 \begin{bmatrix} \\ \\ \end{bmatrix} \oplus a_3 [] \oplus \ldots \oplus a_n \begin{bmatrix} \\ \end{bmatrix}.$$

Unitary group representations   The matrices $A \in \{M_n\left(C\right)\}$ (where $\{M_n\left(C\right)\}$ is a set of square matrices of the n-th order over a field of complex numbers C), such as for any $\bar{a}, \bar{b} \in V$ (V vector space) the following equation is satisfied, $\left(\bar{a}, \bar{b}\right) = \left(A\bar{a}, A\bar{b}\right)$, where $(\,,\,)$ means the scalar product, are called unitary matrices.

Unitary matrices A satisfy the relationship,

$$AA^+ = A^+A = 1 \Rightarrow A^+ = A^{-1},$$

where $A^+ = \left(A^T\right)^*$ and creates a unitary group $U\left(n\right)$.

A unitary representation is a representation made of unitary matrices.

Every representation of a finite group is equivalent to a unitary representation, i.e.,

$$\underset{\{D(x)\}}{\forall}\ \underset{S}{\exists}\ \underset{x \in G}{\forall}\ \underset{\bar{a}, \bar{b} \in V}{\forall}\ \left(D'\left(x\right)\bar{a}, D'\left(x\right)\bar{b}\right) = \left(\bar{a}, \bar{b}\right),$$

where S means nonsingular matrix and $D'\left(x\right) = S^{-1}D\left(x\right)S$.

7.4 Homotopy groups

The arcwise connected space Ω is a space in which each pair of points can be joined within this space with an arc. The arc is understood as homeomorphic with a section $[0,1]$, where this homeomorphism is a transformation $f : X \to Y$ that is continuous and injective, and the reverse transformation $f^{-1} : f(X) \to X$ is continuous as well. We may define the loop (continuous images for the circle S^1 in a given space) in the point ω_0 in the space Ω, as a closed, continuous curved line (with a defined line direction) that passes through the ω_0 point. Such a definition of loops may be treated as continuous transformations $F(t)$ of the section $[0,1]$ into the space Ω, assuming that $F(0) = F(1) = \omega_0$. If $F(t) = \omega_0$ for every $t \in [0,1]$, then the loop is called a zero loop (a point).

It is worth noting that algebraic topology is not only limited to studying topological spaces and the properties of their individual transformations, but also concerns the studies of entire classes of these transformations [Bloom (1979); Birkhoff and Lane (1977); Hatcher (2002); Lang (2002); Mermin (1979); Spanier (1966)]. The notion of homotopy plays a major role in operating these classes. Homotopy can be demonstrated as follows [Bloom (1979); Birkhoff and Lane (1977); Hatcher (2002); Lang (2002); Mermin (1979); Spanier (1966)]: two transformations $f, g : X \to Y$ are homotopic if it is possible to continuously transform the chart of the function f to the chart of the function g in the Cartesian product $X \times Y$. In other words, if there is a family of transformations, $h_t : X \to Y$, which is continuous with respect to the $t \in [0,1]$ parameter, while the first element of this family is $h_0 = f$ and the last one is $h_1 = g$. For example, each transformation of the section $[0,1]$ into itself can be deformed by homotopy to another transformation, in particular, to the constant transformation $c : [0,1] \to [0,1] : x \mapsto p$. One may observe that homotopy divides the vast set of transformations of one space into another one onto equivalence classes, wherein the transformations belonging to the same class have very similar topological properties. For some pairs of spaces, the set of homotopy classes has a natural structure of an algebraic group. The notion of homotopy has become the basis of algebraic topology (particularly the topology of singularities), because, nearly all algebraic invariants are homotopy invariants [Mermin (1979); Spanier (1966)].

7.4.1   Definition of homotopy

The family of transformations

$$h_t : X \to Y, \quad t \in [0,1]$$

is called the homotopy between transformations h_0 a h_1, when the mapping

$$H : X \times [0,1] \to Y : (x,t) \mapsto h_t(x)$$

is continuous.

7.4.2   Homotopic transformations

Two transformations $f : X \to Y$ and $g : X \to Y$ are called homotopic when
there is a homotopy $h_t : X \to Y$, $t \in [0,1]$, such that $h_0 = f$, $h_1 = g$.

Homotopy h_t or H between f and g is denoted by the symbol $h_t : f \sim g$
or $H : f \sim g$.

7.4.3   Properties of homotopy

Suppose X and Y are topological spaces[7]. The homotopic relation in the
set Y^X (of all transformations of the space X into the space Y) is a relation
of equivalence (i.e., it is reflexive, symmetric and transitive). In order to
confirm the reflexivity of the relationship, we just need to assume $\underset{t \in [0,1]}{\forall} h_t =$
f. To confirm its symmetry, we need to assume that $h_t : f \sim g$ is a
homotopy between f and g, and, thus, the homotopy $h_{1-t} : g \sim f$ exists.
In order to confirm the transitivity of the relation, we need to consider two
homotopies: $\phi_t : f \sim g$ and $\psi_t : g \sim h$; then, homotopy $h_t : f \sim h$ is
described with the formula:

$$h_t = \begin{cases} \phi_{2t}, & t \in \left[0, \frac{1}{2}\right] \\ \psi_{2t-1}, & t \in \left[\frac{1}{2}, 1\right] \end{cases}.$$

As demonstrated above, the homotopy relationship is the relationship of
equivalence; therefore, there exist classes of equivalence, called homotopy
classes. The homotopy class, to which the transformation f belongs, is
called the homotopy class of transformation f and is denoted by the symbol
$[f]$.

Let X be a metrizable space. In compact-open topology in Y^X, homotopy classes are identical to the elements of arcwise connection.

---

[7]A topological space is a space X with a defined topology τ that meets the following
conditions: $\emptyset \in \tau$, $X \in \tau$; $\underset{U_1, U_2 \in \tau}{\forall} U_1 \cap U_2 \in \tau$; $\underset{U_s \in \tau}{\forall} \underset{s \in S}{\forall} \bigcup_{s \in S} U_s \in \tau$. This means
that topology is a family of open sets.

7.4.4 *Loop homotopy*

Two loops, F and G, are homotopic if there is a family of transformations $h_t(l)$, $t, l \in [0, 1]$ into a given space in which the loops ($h_t(l)$ are continuous with respect to t and l) and where the following conditions are fulfilled,

$$h_0(l) = F(l),$$

$$h_1(l) = G(l),$$

$$h_t(0) = h_t(1) = \omega_0,$$

where $t, l \in [0, 1]$.

Examples of homotopic and nonhomotopic loop are presented in Fig. 7.2.

a) b)

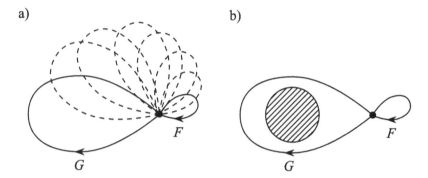

Fig. 7.2 A space with a hole (hatched area). a) Loops F and G are homotopic; the schematic continuous transformation of the F loop into the G loop is presented (both loops are also homotopic with the zero loop); b) Loops F and G are nonhomotopic and lack the possibility of a continuous transformation of the F loop (homotopic with the zero loop) into the G loop (surrounding the hole in space).

The arcwise connected space, in which every loop is homotopic with the zero loop, is called a simply connected space; in other words, a simply connected space X is an arcwise connected space for which every transformation of the $S^1 \to X$ circle is homotopic with a constant transformation or is any space in which every loop is the edge of a 2D sphere image.

We may consider composing two loops after taking into account their directions. The result of composing the paths F and G is referred to as first describing the F loop and then the G loop in the following way:

$$F \cdot G = \begin{cases} F(2t) & t \in [0, \frac{1}{2}] \\ G(2t-1) & t \in [\frac{1}{2}, 1] \end{cases}.$$

It needs to be remembered that the direction of the loop must be considered when describing the loop.

The abstract class for the F loop is denoted as $[F]$, a name for a set with all loops homotopic with the F loop. The set of loops passing through the point ω_0 in a given space Ω can be divided into classes, including those of mutually homotopic loops. Each loop belonging to the class $[F]$ is its representative.

Operations carried out on classes are also defined in a similar way to actions carried out on individual loops [Hatcher (2002); Mermin (1979); Spanier (1966)]:

$$[F] \cdot [G] \overset{def}{=} [F \cdot G].$$

It is important to note that the result of $[F \cdot G]$ composition of two classes of homotopic loops does not depend on individual representatives of these classes $[F]$ and $[G]$. Such a definition of operation results from the fact that any two loops, F and G, attached to the point ω_0 can be composed to produce the $F \cdot G$ loop, which is also attached to the point ω_0. Therefore, we may suggest similar operations on classes that will not exceed the set of these classes. It is shown that the above classes make a group in the given definition of group action.

Suppose $\pi_1(\Omega, \omega_0)$ is a set of classes of homotopic loops in the space Ω attached to the ω_0 point with the operation of loop composition described above.

The associativity of the operation on loop classes needs proving that, if $[F], [G], [H] \in \pi_1(\Omega, \omega_0)$, then

$$[F] \cdot ([G] \cdot [H]) = ([F] \cdot [G]) \cdot [H].$$

The product $[F] \cdot ([G] \cdot [H])$ is a class of homotopic loops,

$$F \cdot (G \cdot H) = \begin{cases} F(4t) & t \in \left[0, \frac{1}{4}\right] \\ G(4t-1) & t \in \left[\frac{1}{4}, \frac{1}{2}\right] \\ H(2t-1) & t \in \left[\frac{1}{2}, 1\right] \end{cases},$$

and the product $([F] \cdot [G]) \cdot [H]$ is also a class of homotopy loops,

$$F \cdot (G \cdot H) = \begin{cases} F(2t) & t \in \left[0, \frac{1}{2}\right] \\ G(4t-2) & t \in \left[\frac{1}{2}, \frac{3}{4}\right] \\ H(4t-3) & t \in \left[\frac{3}{4}, 1\right] \end{cases}.$$

Both classes are homotopic; the homotopy between them is defined by the family of transformations,

$$h_s = \begin{cases} F\left(\frac{4t}{s+1}\right) & t \in \left[0, \frac{s+1}{4}\right] \\ G(4t-s-1) & t \in \left[\frac{s+1}{4}, \frac{s+2}{4}\right] \\ H\left(\frac{4t-s-2}{2-s}\right) & t \in \left[\frac{s+2}{4}, 1\right] \end{cases},$$

which proves that the operation is associative.

The neutral element of the group $\pi_1\left(\Omega, \omega_0\right)$ is the class of loops homotopic with a zero loop (a point). If $[F] \in \pi_1\left(\Omega, \omega_0\right)$, then

$$[F] \cdot [\varepsilon] = [F] \qquad [\varepsilon] \cdot [F] = [F].$$

The homotopy between the loops $F \cdot \varepsilon$ and F is defined by the family of relations,

$$h_s = \begin{cases} F\left(\frac{2t}{1+s}\right) & t \in \left[0, \frac{s+1}{2}\right] \\ \omega_0 & t \in \left[\frac{s+1}{2}, 1\right] \end{cases},$$

and the homotopy between the loops $\varepsilon \cdot F$ and F is defined by the family of transformations,

$$h_s = \begin{cases} \omega_0 & t \in \left[0, \frac{1-s}{2}\right] \\ F\left(\frac{2t+s-1}{1+s}\right) & t \in \left[\frac{1-s}{2}, 1\right] \end{cases},$$

which indicates the existence of a neutral element.

The inverse element relative to the class $[F] \in \pi_1\left(\Omega, \omega_0\right)$ is the class that includes the inverse loop F^{-1}, i.e., $[F]^{-1} = \left[F^{-1}\right]$. For every class $[F] \in \pi_1\left(\Omega, \omega_0\right)$,

$$[F] \cdot [F]^{-1} = [e] = [F]^{-1} \cdot [F].$$

The homotopy between the loops $F \cdot F^{-1}$ and e is defined by the family of transformations,

$$h_s = \begin{cases} F\left(2t\right) & t \in \left[0, \frac{1-s}{2}\right] \\ F\left(2 - 2\left(t+s\right)\right) & t \in \left[\frac{1-s}{2}, 1-s\right] \\ F\left(0\right) & t \in \left[1-s, 1\right] \end{cases}.$$

For $s = 0$, $h_0 = F \cdot F^{-1}$ and for $s = 1$, $h_1 = e$, therefore, $[F] \cdot [F]^{-1} = \left[F \cdot F^{-1}\right] = [e]$. The proof for $[F]^{-1} \cdot [F] = [e]$ takes an analogical course.

The group described above is called the fundamental (basic) group of the Ω space in the ω_0 point, or **the first homotopy group** [Bloom (1979); Birkhoff and Lane (1977); Hatcher (2002); Lang (2002); Mermin (1979); Spanier (1966)],

$$\pi_1\left(\Omega, \omega_0\right).$$

If the Ω space is a simple-connected space, the first homotopy group for this space is a trivial group, i.e., it only includes one element, $\pi_1\left(\Omega, \omega_0\right) = \varepsilon$ (where ε is a single-element group–a trivial group). The distinguished point ω_0 in the Ω space is called the base point. The class of zero loop includes an identity element of the group $\pi_1(\Omega, \omega_0)$. The reverse element of the loop class $[F]$ is defined as follows:

$$[F]^{-1} = [F^{-1}],$$

where F^{-1} is a loop directed opposite to the F loop.

We may now consider the C curve, which belongs to the arcwise connected space Ω and joins the points ω_0 and ω_1 in this space. The below transformation:

$$\pi_1(\Omega, \omega_0) \to \pi_1(\Omega, \omega_1) : [F] \to [C] \cdot [F] \cdot [C]^{-1}, \tag{7.22}$$

is an isomorphism, i.e., a one-to-one homomorphism (the above isomorphism is presented in Fig. 7.3).

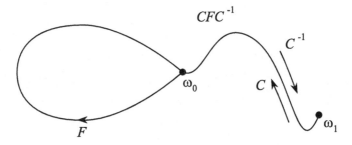

Fig. 7.3   Construction of the loop $C \cdot F \cdot C^{-1}$ in the point ω_1.

One can demonstrate that the projection $\pi_1(\Omega, \omega_0) \to \pi_1(\Omega, \omega_1)$: $[F] \to [C] \cdot [F] \cdot [C]^{-1}$ is an isomorphism. If $[F] \in \pi_1(\Omega, \omega_0)$, then the following composition holds: $[C] \cdot [F] \cdot [C]^{-1} \in \pi_1(\Omega, \omega_1)$, and therefore, the image (7.22) is properly defined. If $[F_1]$, $[F_2] \in \pi_1(\Omega, \omega_0)$, then the composition $[F_1] \cdot [F_2]$ corresponds to the element,

$$[C] \cdot ([F_1] \cdot [F_2]) \cdot [C]^{-1} = \left([C] \cdot [F_1] \cdot [C]^{-1}\right) \cdot \left([C] \cdot [F_2] \cdot [C]^{-1}\right),$$

thus, the projection (7.22) is an isomorphism.

If $[C] \cdot [F] \cdot [C]^{-1} = e$ (neutral element within the group $\pi_1(\Omega, \omega_1)$), then

$$[C]^{-1} \cdot [C] \cdot [F] \cdot [C]^{-1} \cdot [C] = [C]^{-1} \cdot [C] = e,$$

(neutral element within the group $\pi_1(\Omega, \omega_0)$); therefore, the projection (7.22) is injective.

If $[G] \in \pi_1(\Omega, \omega_1)$, then $[C]^{-1} \cdot [F] \cdot [C] \in \pi_1(\Omega, \omega_0)$ and

$$[C]^{-1} \cdot [F] \cdot [C] \to [C],$$

and thus, the projection (7.22) is surjective ("onto").

If the projection is injective and surjective, then it is one to one, i.e., it is an isomorphism.

The above suggests that the fundamental group (the first homotopy group) for an arcwise connected space does not depend on the choice of the base point. For that reason, in arcwise connected spaces, the existence of the fundamental group, i.e., the homotopy group, does not require defining the base point (the first homotopy group can therefore be written as $\pi_1(\Omega)$).

Below, we will present considerations that allow to determine fundamental groups for some more complex spaces. The fundamental group for the Cartesian product of two arcwise connected spaces equals to the direct sum of two fundamental groups for these spaces [Hatcher (2002)],

$$\pi_1(\Omega_1 \times \Omega_2) = \pi_1(\Omega_1) \oplus \pi_1(\Omega_2),$$

where the direct sum of $G_1 \oplus \ldots \oplus G_n$ groups $G_1, \ldots, G_n$ is the Cartesian product $G_1 \times \ldots \times G_n$ with the operation " $\bullet$ " determined with the formula,

$$(a_1, \ldots, a_n) \bullet (b_1, \ldots, b_n) = \left(a_1 \overset{1}{\bullet} b_1, \ldots, a_n \overset{n}{\bullet} b_n \right),$$

where " $\overset{k}{\bullet}$ " is an operation (action) within the group G_k, $k = 1, 2, \ldots, n$.

The real space R is simply connected, thus,

$$\pi_1(R) = \varepsilon.$$

So also:

$$\pi_1(R^n) = \varepsilon.$$

(where ε is a single-element group–a trivial group) because

$$\pi_1(R^n) = \pi_1\left(R^{n-1} \times R\right) = \ldots = \pi_1\left(\underbrace{R \times \ldots \times R}_{n}\right) = \underbrace{\pi_1(R) \oplus \ldots \oplus \pi_1(R)}_{n}.$$

We may now consider $\pi_1(S^1)$. The elements of the fundamental group in an arcwise connected space S^1 are the classes with a given multiplicity of directed loops (the number of directed windings over the circle); therefore, it is a group isomorphic with an additive group of integers Z. Integers correspond to the number of windings, and the integer adding operation corresponds to composing loops. Therefore, knowing that,

$$\pi_1(S^1) = Z,$$

similarly, as in the case of R^n, we may characterize a homotopy group for a torus as $T = S^1 \times S^1$, namely,

$$\pi_1 (T) = \pi_1 \left(S^1 \times S^1 \right) = \pi_1 \left(S^1 \right) \oplus \pi_1 \left(S^1 \right) = Z \oplus Z.$$

We may then characterize higher homotopy groups analogously. According to the definition, a loop is a transformation of f section $[0, 1]$ into a space Ω, assuming that $f(0) = f(1) = \omega_0$. Loops can also be obtained as transformations of circles instead of sections, i.e.,

$$f : S^1, s \to \Omega, \omega_0,$$

circle S^1 is transformed into Ω, and the point $s \in S^1$ passes into its image $\omega_0 \in \Omega$. We may treat it as the definition of the fundamental group; the fundamental group is a set of homotopy classes of transformations $f : S^1, s \to \Omega, \omega_0$. We now may naturally introduce the n-th homotopy group (n-dimensional homotopy group) as a set of homotopy classes of transformation $f : S^n, s \to \Omega, \omega_0$, for $n = 0, 1, 2, \ldots$. These groups will be denoted as,

$$\pi_n (\Omega, \omega_0).$$

The above definition is correct for $n > 0$. When $n = 0$, the set $\pi_0 (\Omega, \omega_0)$ obtained from the definition does not have to be a group because $\pi_0 (\Omega, \omega_0)$ is a set of arc elements, i.e., the maximal arcwise connected subspaces of a given space, of the Ω space with the ω_0 point as its base point. For the arcwise connected space, $\pi_0 (\Omega, \omega_0)$ is a single-element group. In some circumstances, we can define the operation for the zero homotopy group and create the group structure [Mermin (1979)].

7.5 Configuration space

The configuration space, or the particle position space is a space, in which, the classical positions of all particles in the system can be characterized depending on the number of degrees of freedom for that multi-particle system. Individual particles can be placed on a particular manifold M, where the manifold is understood as the space that has a local structure of the Euclidean space, e.g., plane R^2, circle S^1, sphere S^2, torus T, etc.

If N identical particles are considered as a set of distinguishable particles, i.e., as particles in the classical sense (they can be enumerated), then the configuration space of the whole system has the form of a N-times

Cartesian product of the manifold M,

$$\underbrace{M \times \ldots \times M}_{N} = M^{N}. \qquad (7.23)$$

Whereas, if the particles are treated as particles in the quantum sense, then one should assume the indistinguishability of identical particles, and the configuration space for a system of such particles will be different such that points from the M^{N} space only differing in the permutation of indices are to be considered identical. To account for the indistinguishability of identical particles, the notion of quotient space is used [8].

The configuration space of N indistinguishable, identical particles on a given manifold M can be written as a quotient space,

$$M^{N}/S_{N}, \qquad (7.24)$$

where S_{N} is a permutation group of N elements.

In the case of spaces with indistinguishable particles M^{N}/S_{N}, it is worth noting the issue of singular points, i.e., such points, for which the position of two or more particles overlap (these are fixed points for operations in the group S_{N} on M^{N} – these points are called diagonal points). The diagonal points (a discrete set of diagonal points is denoted as Δ) must be removed from the space in order to ensure maintaining a constant number of particles in the system. In other words, the configuration space of the system of N identical, indistinguishable particles on the manifold M has the form:

$$Q_{N}(M) = \left(M^{N} \backslash \Delta\right)/S_{N}. \qquad (7.25)$$

Similarly, the configuration space of a system of N identical, distinguishable particles on the manifold M, after removal of the diagonal points, attains the form:

$$F_{N}(M) = M^{N} \backslash \Delta. \qquad (7.26)$$

---

[8]If X is a linear space and Y is its linear subset, then abstract classes are the subsets of space X obtained in such a way that x and y belong to the same abstract class, when and only when $x - y \in Y$. The class, which includes the x element, is usually denoted as $[x]$. In the set of abstract classes, we can introduce the operations $[x] + [y] = [x + y]$ and $\alpha[x] = [\alpha x]$, where α is a number (these operations meet the axioms of linear space operations). The zero element is the abstract class to which all elements from the set Y belong. The set of all abstract classes together with this definition of operations is called the *quotient space* and is denoted as X/Y.

7.5.1   *First homotopy group of configuration space for many particle systems*

The first homotopy group for the configuration space of a system of many particles on a given manifold is called a braid group [Birman (1974); Jacak *et al.* (2003)]. If we consider the configuration space of a system of N identical, indistinguishable particles on the manifold M, then the first homotopy group for this space is called a full homotopy group and is written in the following way,

$$\pi_1 \left(\left(M^N \backslash \Delta \right) / S_N \right). \tag{7.27}$$

Elements belonging to this group consist of homotopic trajectories (i.e., trajectories that may be topologically transformed into one another) in the quotient space $\left(M^N \backslash \Delta \right) / S_N$. In this case, the final and initial positions can differ; they are related to each other by a permutation of particle coordinates (thus, they are indistinguishable).

Whereas, for the configuration space of a system of N identical but distinguishable particles on the manifold M, the first homotopy group is called the pure braid group,

$$\pi_1 \left(M^N \backslash \Delta \right). \tag{7.28}$$

Elements building the pure braid group are classes consisting of homotopic trajectories in the space $M^N \backslash \Delta$ for which the final and initial positions are the same (also with regard to the particle ordering).

A braid group can be presented using its generators[9]. In the case of a pure braid group, the generators are topological classes of non-contractible[10] (nonhomotopic with the point) loops corresponding to such a trajectory in the system when only one particle traverses its trajectory and the rest remain in place. This single loop is a non-contractible, closed trajectory–its non-contractibility is caused by one of two factors:

(1) the remaining particles (in some cases, depending on the type and size of the manifold, the remaining particles are responsible for the existence of homotopy classes),

(2) topological defects in the considered manifold M.

---

[9]The generator of the G group is the element from the X set that satisfies the following conditions: 1) every element in the group G can be presented as a product of the elements of the X set or their inverse elements and 2) the smallest subgroup of the G group, which includes the X set, is the G group.

[10]Contractibility means a continuous transformation in a topological sense, from a loop to a point (an example of a non-continuous transformation is cutting the loop).

The full braid group $\pi_1\left(\left(M^N\backslash\Delta\right)/S_N\right)$ is therefore generated by elements, which can be divided into two types. The first type corresponds to generators of the so-called pure braid group [Birman (1974); Jacak *et al.* (2003)], and the second represents classes of homotopic, closed trajectories corresponding to the exchange between two particles (with a lack of changes in the location of the other particles and a lack of additional non-homotopic loops). These two types of generators that generate the group $\pi_1\left(\left(M^N\backslash\Delta\right)/S_N\right)$ are denoted as L and P, respectively. We may notice that L is a set of generators of the group $\pi_1\left(M^N\backslash\Delta\right)$, whereas the group generated by the set P is denoted by $\Sigma_N(M)$. For a simply connected manifold M [Birman (1974); Jacak *et al.* (2003)],

$$\pi_1\left(\left(M^N\backslash\Delta\right)/S_N\right) = \Sigma_N(M) \tag{7.29}$$

and the group $\pi_1\left(M^N\backslash\Delta\right)$ is a subgroup of the $\Sigma_N(M)$ [Birman (1974)]. This expression results from the fact that for the simple connected manifold M generates the full braid group $\pi_1\left(\left(M^N\backslash\Delta\right)/S_N\right)$ (and all elements of the set L) using only generators from the set P. If the manifold M is not simply connected, then the sets L and P are separable and the relationships between their elements depend on the manifold M [Birman (1974); Imbo *et al.* (1990)]. We may observe that the group $\pi_1\left(M^N\backslash\Delta\right)$ is a normal subgroup $\pi_1\left(\left(M^N\backslash\Delta\right)/S_N\right)$ for any manifold M. Elements of the quotient group $\pi_1\left(\left(M^N\backslash\Delta\right)/S_N\right)/\pi_1\left(M^N\backslash\Delta\right)$ number all possible particle permutations that can be obtained from any initial ordering of the particles. The quotient group,

$$\pi_1\left(\left(M^N\backslash\Delta\right)/S_N\right)/\pi_1\left(M^N\backslash\Delta\right),$$

is generally a permutation group S_N, but there are manifolds for which that is not the case [Birman (1974); Imbo *et al.* (1990)], e.g., for compact manifolds, including the circle [Birman (1974); Jacak *et al.* (2003)]. In the case of many particles on a compact manifold M, which is the sum of maximal compact spaces,

$$M = \sum_{i=1}^{k} M_i, \tag{7.30}$$

all possible arrangements of the system will be described by elements of the following group,

$$\tilde{S}_N = \bigoplus_{i=1}^{k} S_{N_i}, \tag{7.31}$$

where N_i means the number of particles in the subspace N_i.

The space $F_N(M) = M^N \backslash \Delta$ is a fibration [11] over the space $Q_N(M) = (M^N \backslash \Delta) / S_N$ imposed by epimorphism [12] $h_{S_N} : F_N(M) \rightarrow Q_N(M)$ [Hatcher (2002); Spanier (1966)]. The linear projection h_{S_N} is related to the operation of the permutation group S_N on the space $F_N(M)$.

In the above case, the fibers [13] are $N!$-element spaces $[F_N(M)]_x$ where $x \in Q_N(M)$. Moreover, it is a locally trivial fibration. The local triviality of this fibration is defined as follows [Birman (1974); Hatcher (2002); Spanier (1966)]:

$$\underset{x \in Q_N(M)}{\forall} \underset{U}{\exists} F_N(M) \,|\, U = U \times [F_N(M)]_x , \qquad (7.32)$$

where U means the surrounding of the point x and $F_N(M) \,|\, U$ means cutting space $F_N(M)$ to the inverse image U in the projection $h_{S_N} : F_N(M) \rightarrow Q_N(M)$.

7.5.2   *Covering space*

The covering of a simply connected space X is defined as a transformation $p : Y \rightarrow X$ when,

$$\underset{x \in X}{\forall} \underset{U}{\exists} p^{-1}(U) = \bigcup_{i \in I} C_i, \quad C_j \cap C_k = \emptyset, \quad int C_i = C_i, \qquad (7.33)$$

where U means an open and arcwise connected surrounding of the point x and when,

$$\underset{C_i}{\forall} p \,|\, C_i : C_i \rightarrow U, \qquad (7.34)$$

where, $p \,|\, C_i$ is a (surjective) homomorphism between C_i and U. If $p : Y \rightarrow X$ is a covering, then Y is called the covering space of the space X. The above statements suggest that the space $F_N(M)$ covers the space $Q_N(M)$. Excluding the singularity points (a set of diagonal points Δ) leads to free operation [14] S_N on $F_N(M)$ indicating that the homomorphism $h_{S_N} : F_N(M) \rightarrow Q_N(M)$ is a covering projection, or a locally trivial projection, and that every fiber $[F_N(M)]_x$ is discrete (contains a finite number of elements) [Birman (1974); Jacak *et al.* (2003)].

---

[11] The fibration on the topological space X is the pair composed of the topological space Y and the continuous mapping $h : Y \rightarrow X$ (surjective transformation).

[12] Epimorphism is a surjective homomorphism ("onto").

[13] A fiber is an inverse image of the point $x \in X$ in a projection $h : Y \rightarrow X$ (or $h^{-1}(x)$).

[14] Free operation is defined as the operation without fixed points, i.e., such points p belonging to a given manifold M that, when the group G operates on M, then $g \in G$, $gp = p$.

7.6 Braid groups for the chosen manifolds

7.6.1 *Braid group for a two-dimensional Euclidean space* R^2

The full braid group for R^2,

$$\pi_1 \left(Q_N \left(R^2 \right) \right) = \pi_1 \left(\left(\left(R^2 \right)^N \backslash \Delta \right) / S_N \right), \qquad (7.35)$$

has been described by Artin [Artin (1947)] (it is called the classical Artin homotopy group) and is often denoted as B_N. For manifolds R^2, a graphical presentation of the elements of a braid group is used as a standard–using the so-called geometric braids. The trivial element of the full braid group for the system of N particles on the manifolds R^2 is represented by nonintersecting lines (imaging trajectories of individual particles) joining N initial points with N final points (Fig. 7.4).

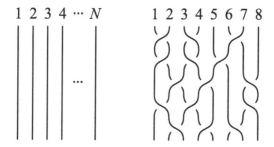

Fig. 7.4 Trivial N-braid and non-trivial 8-braid

It is significant that the lines in the figure representing individual particle trajectories in the non-trivial case may generate braids through any form of entanglement, provided the lack of possibility to intercross individual lines, because diagonal points have been excluded from the space in question (Δ, the set of diagonal points was removed) (Fig. 7.4).

If the initial ordering of particles is identical to the final ordering and the braid describes a closed curve (loop) in the space $F_N \left(R^2 \right) = \left(R^2 \right)^N \backslash \Delta$, then such a braid cannot be disentangled because it would require 'cutting' the braid line.

In the view of the fact that R^2 is a simply connected space, one can write,

$$\pi_1 \left(R^2 \right) = \varepsilon, \qquad (7.36)$$

and the full braid group B_N is generated by generators from the set P, which are elements corresponding to the exchange of two neighboring particles. The element σ_i corresponding to the exchange of the i-th particle with the $i + 1$-th particle and its inverse element σ_i^{-1} are presented in Fig. 7.5.

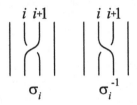

Fig. 7.5   Graphical presentation of the elementary operations of the particle exchange σ_i and the inverse exchange σ_i^{-1}.

Therefore, one may state that,

$$B_N = \Sigma_N \left(R^2 \right). \tag{7.37}$$

It is also important to present the relationships that determine the generators σ_i of the full braid group [Birman (1974)].

The condition imposed on generators σ_i of B_N Artin's group,

$$\sigma_i \cdot \sigma_{i+1} \cdot \sigma_i = \sigma_{i+1} \cdot \sigma_i \cdot \sigma_{i+1}, \qquad 1 \leq i \leq N - 2, \tag{7.38}$$

is related to the possibility of 'shifting' the braid (graphical presentation of this relation is depicted in Fig. 7.6).

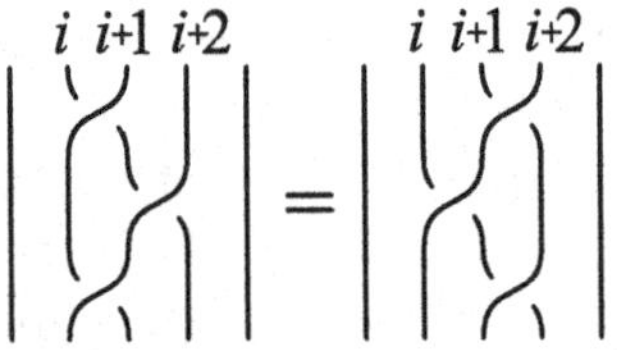

Fig. 7.6   The left-hand braid is topologically homotopic with the right-hand braid (the transformation between one and the other is topologically continuous).

Another condition imposed on generators σ_i,

$$\sigma_i \cdot \sigma_j = \sigma_j \cdot \sigma_i, \qquad 1 \leq i, j \leq N - 1, |i - j| \geq 2, \tag{7.39}$$

describes the lack of impact of the succession of particle exchange (i-th particle with $i+1$-th and particle j-th with particle $j+1$-th) on the topology of braids, provided that the particular exchanges do not overlap, which is

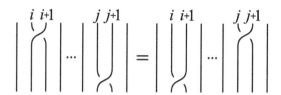

Fig. 7.7 The left-hand braid is topologically equivalent to the right-hand braid (the succession of the non-overlapping exchange of particle pairs does not affect loop homotopy).

guaranteed by the condition $|i - j| \geq 2$ (this condition is illustrated in Fig. 7.7).

It is worth emphasizing that the operation of a double particle exchange σ_i^2 (presented in Fig. 7.8) is not topologically equivalent to the trivial braid because individual lines i-th and $i+1$-th hook each other, making an entangled braid.

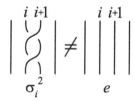

Fig. 7.8 A double exchange of the particles i-th and $i+1$-th denoted as σ_i^2 is not a unit element of the group B_N.

Although the double exchange σ_i^2 maintains the ordering of particles,

$$\sigma_i^2 \neq e. \tag{7.40}$$

In the group B_N, not only are the initial and final configurations of a given system important (given points in the space $F_N(M) = M^N \backslash \Delta$) but the very course of trajectory in the system is also important.

Considering a set of generators L for the manifold R^2, one may observe the first homotopy group for the configuration space of N particles on the manifold R^2 is a subgroup of the classical Artin homotopy group,

$$\pi_1\left(F_N\left(R^2\right)\right) \subset \Sigma_N\left(R^2\right). \tag{7.41}$$

This result suggests that the generators from the set L can be presented using elements σ_i,

$$l_{ij} = \sigma_{j-1} \cdot \sigma_{j-2} \cdot \ldots \cdot \sigma_{i+1} \cdot \sigma_i^2 \cdot \sigma_{i+1}^{-1} \cdot \ldots \cdot \sigma_{j-2}^{-1} \cdot \sigma_{j-1}^{-1}, \quad 1 \leq i \leq j \leq N-1 \tag{7.42}$$

Element l_{ij}, or the element of the pure braid group, which is also its generator, corresponds to the exchange of the i-th particle with the j-th particle, preserving the ordering of the particles (Fig. 7.9).

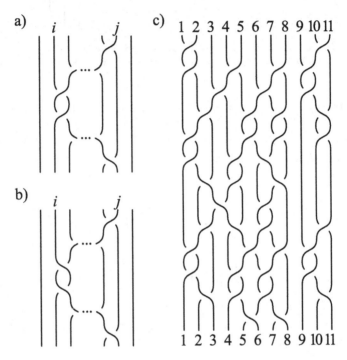

Fig. 7.9   a) The generator l_{ij} of the pure braid group corresponds to the exchange of the i-th and j-th particles while preserving the particle ordering. b) The inverse element $l_{ij}^{-1} = \sigma_{j-1} \cdot \sigma_{j-2} \cdot \cdots \cdot \sigma_{i+1} \cdot \sigma_i^{-2} \cdot \sigma_{i+1}^{-1} \cdot \cdots \cdot \sigma_{j-2}^{-1} \cdot \sigma_{j-1}^{-1}$ to the generator l_{ij}. c) Exemplary 11-braid, which is an element of the pure braid group; the initial and final particle positions are identical.

The relationships imposed on the generators define the pure braid group $\pi_1\left(\left(R^2\right)^N \backslash \Delta\right)$ [Coxeter and Moser (1984)]

$$
l_{rs}^{-1} \cdot l_{ij} \cdot l_{rs} = \begin{cases} l_{ij}, & i < r < s < j \\ l_{ij}, & r < s < i < j \\ l_{rj} \cdot l_{ij} \cdot l_{rj}^{-1}, & r < i = s < j \\ l_{rj} \cdot l_{sj} \cdot l_{ij} \cdot l_{sj}^{-1} \cdot l_{rj}^{-1}, & i = r < s < j \\ l_{rj} \cdot l_{sj} \cdot l_{rj}^{-1} \cdot l_{sj}^{-1} \cdot l_{ij} \cdot l_{sj} \cdot l_{rj} \cdot l_{sj}^{-1} \cdot l_{rj}^{-1}, & r < i < s < j \end{cases}
$$

$$(7.43)$$

We may now move onto the quotient group,

$$\pi_1\left(\left(\left(R^2\right)^N\backslash\Delta\right)/S_N\right)/\pi_1\left(\left(R^2\right)^N\backslash\Delta\right) = B_N/\pi_1\left(F_N\left(R^2\right)\right). \quad (7.44)$$

According to the definition, a quotient structure is represented using the homomorphism, with the kernel being a normal subgroup of the pure braid group $\pi_1\left(\left(R^2\right)^N\backslash\Delta\right)$. In other words, it is determined by the condition imposed on the group generators,

$$l_{ij} = e, \quad (7.45)$$

or,

$$l_{ij} = \sigma_{j-1}\cdot\sigma_{j-2}\cdot\ldots\cdot\sigma_{i+1}\cdot\sigma_i^2\cdot\sigma_{i+1}^{-1}\cdot\ldots\cdot\sigma_{j-2}^{-1}\cdot\sigma_{j-1}^{-1} = e, \quad (7.46)$$

where $1 \le i \le j \le N - 1$, therefore,

$$\sigma_i^2 = e, \quad (7.47)$$

where $1 \le i \le N - 1$.

The above expression indicates that the quotient group is generated by the generators σ_i, which satisfy the relationships (7.38) and (7.39) as well as the relationship (7.47). A group generated in this way turns out to be a permutation group of the N-element set, denoted as S_N. In other words,

$$\pi_1\left(\left(\left(R^2\right)^N\backslash\Delta\right)/S_N\right)/\pi_1\left(\left(R^2\right)^N\backslash\Delta\right) = B_N/\pi_1\left(F_N\left(R^2\right)\right) = S_N. \quad (7.48)$$

The permutation group S_N of a N-element set is generated by exchange operations between neighboring elements of the set, i.e., it is created by generators, which are analogous to σ_i generators. The generators of the group S_N must therefore satisfy the relationships determining the generators σ_i, i.e., (7.38) and (7.39) (Fig. 7.6 and Fig. 7.7). In contrast to the group B_N, which expresses a dependence not only on the configuration but also on the course of the trajectory itself, the permutation group S_N depends only on the ordering of the elements and not on the particular way of the exchange. Thus, there is no difference between the element σ_i^2 and the unit element e in the permutation group, which also means that there is no difference between σ_i and σ_i^{-1}. Therefore, generators σ_i satisfy the condition (7.47).

There is also another presentation of the classical Artin homotopy group B_N described in [Birman (1969)] that introduces new generators,

$$\begin{aligned}\sigma &= \sigma_1,\\ a &= \sigma_1\cdot\sigma_2\cdot\ldots\cdot\sigma_{N-1}.\end{aligned} \quad (7.49)$$

The above generators are determined by the following relationships:,

$$a^N = (a \cdot \sigma)^{N-1}$$

and

$$\sigma \cdot a^{-j} \cdot \sigma \cdot a^j = a^{-j} \cdot \sigma \cdot a^j \cdot \sigma,$$

where $2 \leq j \leq \frac{N}{2}$. We can represent generators σ_i with the use of generators σ, a,

$$\sigma_i = a^{i-1} \cdot \sigma \cdot a^{-(i-1)}.$$

In contrast to the previously discussed presentation of B_N, in which the generator corresponded to the exchange operations of two neighboring particles (generators σ_i), the above presentation is based on generators corresponding to the exchange of two neighboring particles but only when their indices are 1 and 2 (generator σ), whereas the compositions (power) of the generator a render other particles to positions 1 and 2.

7.6.2   *Braid group for a sphere S^2*

S^2 is a simply connected space because every loop is homotopic with the zero loop, i.e., the first homotopy group for this manifold has the form,

$$\pi_1\left(S^2\right) = \varepsilon.$$

The braid group for the system of N particles on a sphere S^2 has the form,

$$\pi_1\left(Q_N\left(S^2\right)\right) = \Sigma_N\left(S^2\right).$$

Similar to the case of the space R^2, there is an intuitive, graphical presentation of a braid group (Fig. 7.10).

A braid group on a sphere is generated by generators corresponding to the exchange of pairs of neighboring particles. Because the sphere S^2 is locally isomorphic with a plane R^2, the properties of generators on R^2 (7.38), (7.39) must also be shared by the generators on the sphere. However, we should bear in mind that the sphere S^2 has different global properties than the space R^2 with which it is locally isomorphic. Because of that, any loop on a sphere can be interpreted in two ways. For instance, a particle may describe a loop around all other particles; such a loop is homotopic with the zero loop (Fig. 7.11).

This difference leads to an additional restriction imposed on generators of the braid group on the sphere [Birman (1974); Imbo *et al.* (1990)]:

$$\sigma_1 \cdot \sigma_2 \cdot \ldots \cdot \sigma_{n-2} \cdot \sigma_{n-1}^2 \cdot \sigma_{n-2} \cdot \ldots \cdot \sigma_2 \cdot \sigma_1 = e. \tag{7.50}$$

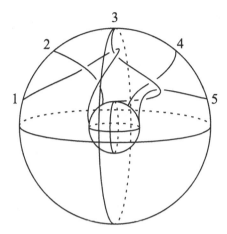

Fig. 7.10 The graphical presentation of the 4-braid on a sphere S^2. Two sets of points (larger sphere and smaller sphere) correspond to the final and initial configurations of a given system on a sphere.

Introducing additional restrictions on generators indicates the poorer character of the braid group on a sphere compared to a braid group on a plane. However, the relationships that determine the generators l_{ij} on a sphere S^2 have an identical form to those on a plane R^2, which is associated with the definition of the generators l_{ij} (7.42) (the restriction (7.50) does not change the relationships for generators of the pure group on a sphere). Therefore, the quotient structures of the full braid group and the pure braid group have the following form:

$$
\begin{aligned}
&\pi_1\left(\left(\left(S^2\right)^N\backslash\Delta\right)/S_N\right)/\pi_1\left(\left(S^2\right)^N\backslash\Delta\right) \\
&= \pi_1\left(Q_N\left(S^2\right)\right)/\pi_1\left(F_N\left(S^2\right)\right) = S_N.
\end{aligned}
\tag{7.51}
$$

7.6.3 *Braid group for a torus T*

A torus is a Cartesian product of two circles S^1:

$$
T = S^1 \times S^1.
$$

As a manifold, the torus has many very important physical interpretations. A torus is topologically equivalent to a square plaquette in the space R^2 with periodical boundary conditions imposed. In contrast to a plane or a sphere, this manifold is not a simply connected space, which means that

$$
\pi_1\left(T\right) \neq \varepsilon.
$$

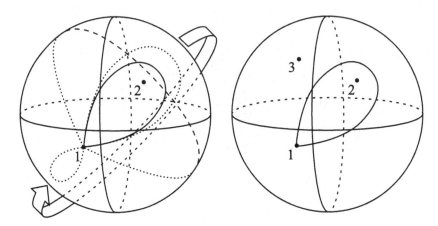

Fig. 7.11   A loop described by a particle around all other particles is homotopic with a point–in the case of two particles on a sphere, each loop is homotopic with a point (left side); in the case of three or more particles, it is impossible to contract the loop (which does not surround all particles) to a point (right side).

The first homotopy group of the system of N particles on the torus T has a more complex structure compared to braid groups for a plane or a sphere [Birman (1974); Einarsson (1990)], which results from the significant topological properties of the torus, i.e., from its multi-connected geometry. The first homotopy group for the torus has the form,

$$\pi_1\left(T\right) = \pi_1\left(S^1 \times S^1\right) = \pi_1\left(S^1\right) \oplus \pi_1\left(S^1\right) = Z \oplus Z \neq \varepsilon,$$

where $\oplus$ means a direct sum and Z is an additive group of integers. The above expression indicates that

$$\pi_1\left(Q_N\left(T\right)\right) \neq \Sigma_N\left(T\right).$$

The set of generators of the group $\pi_1\left(Q_N\left(T\right)\right)$ consists of two sets L and P (because the manifold in question is not simply connected). The set of generators L includes the elements generating the pure braid group $\pi_1\left(F_N\left(T\right)\right)$, and the generators from the set P are responsible for the exchanges of neighboring particles on the surface of the torus. We should note that not all elements of the set L can be presented using generator compositions responsible for particle exchanges. The ones that are not expressed by generators from the set P are related to the topological properties of the torus, and even a single particle can have nonhomotopic trajectories. A graphical presentation of a braid group on the torus enables an easy display of generators belonging to the set L, which are associated with the topology of the torus (Fig. 7.12).

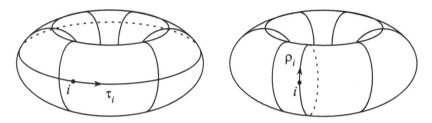

Fig. 7.12 A graphical presentation of generators τ_i and ρ_i not associated with particle exchanges but resulting from the topological properties of the torus.

Generators τ_i and ρ_i (where $i = 1, 2, \ldots, N$) as defined in Fig. 7.12, are responsible for the traversing of a single particle along one of the two types of trajectories on the torus; in both cases, the other particles remain still. For the sake of convenience, new generators associated with particle looping can be introduced. These generators, A_{ij} and C_{ij}, correspond to two types of nonhomotopic trajectories described by the i-th particle around the j-th particle (Fig. 7.13).

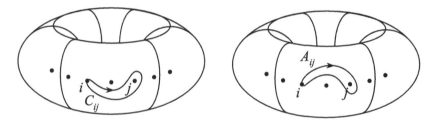

Fig. 7.13 Auxiliary generators A_{ij} and C_{ij} (describing the looping of particles) and the corresponding nonhomotopic trajectories.

New generators, A_{ij} and C_{ij}, can be expressed using generators τ_i and ρ_i through the following relations [Birman (1974); Einarsson (1990)]:

$$A_{ij} = \tau_j^{-1} \cdot \rho_i \cdot \tau_j \cdot \rho_i^{-1},$$
$$C_{ij} = \rho_j^{-1} \cdot \tau_i \cdot \rho_j \cdot \tau_i^{-1}.$$

where $1 \le i \le j \le N$. Thus, we obtain a set of conditions defining the generators of the full braid group for the system of N particles on a torus manifold, which can be divided into three subsets:

(1) equations defining the generators τ_i i ρ_i,
(2) equations defining the generators σ_i,
(3) equations defining the dependencies between generators from the sets L and P.

The below equations define the generators from the set L [Birman (1974); Einarsson (1990)]:

$$\tau_i \cdot A_{jk} = A_{jk} \cdot \tau_i,$$
$$\rho_i \cdot A_{jk} = A_{jk} \cdot \rho_i,$$
$$\tau_i \cdot \tau_j = \tau_j \cdot \tau_i,$$
$$\rho_i \cdot \rho_j = \rho_j \cdot \rho_i,$$
$$C_{ij} = (\tau_i \cdot \tau_j) \cdot A_{ij}^{-1} \cdot \left(\tau_j^{-1} \cdot \tau_i^{-1}\right),$$
$$A_{ij} = (\rho_i \cdot \rho_j) \cdot C_{ij}^{-1} \cdot \left(\rho_j^{-1} \cdot \rho_i^{-1}\right),$$
$$C_{ij} = \left(A_{j-1,j}^{-1} \cdot A_{j-2,j}^{-1} \cdot \ldots \cdot A_{i+1,j}^{-1}\right) \cdot A_{ij}^{-1} \cdot \left(A_{i+1,j} \cdot A_{i+2,j} \cdot \ldots \cdot A_{j-1,j}\right),$$
$$\tau_1 \cdot \rho_1 \cdot \tau_1^{-1} \cdot \rho_1^{-1} = A_{1,N} \cdot A_{1,N-1} \cdot \ldots \cdot A_{1,3} \cdot A_{1,2}.$$

where $1 \leq i \leq j \leq k \leq N$. The equations defining the generators σ_i have an identical form with equations (7.38) and (7.39). The third subset of the relationships defining the generators of the full braid group, which determines the mutual relationships of the generators from the sets L and P, has the form [Birman (1974); Einarsson (1990)]:

$$\tau_{i+1} = \sigma_i^{-1} \cdot \tau_i \cdot \sigma_i^{-1},$$
$$\rho_{i+1} = \sigma_i \cdot \rho_i \cdot \sigma_i,$$
$$\tau_1 \cdot \sigma_i = \sigma_i \cdot \tau_1,$$
$$\rho_1 \cdot \sigma_i = \sigma_i \cdot \rho_1,$$
$$\sigma_i^2 = A_{i,i+1},$$

where $1 \leq i \leq N - 1$ i $2 \leq j \leq N - 1$.

What is important is that the given examples of braid groups for the system of N particles on the Euclidean plane, sphere and torus manifolds fully represent the global topological properties of these spaces [Mermin (1979)]. In the case of the sphere, which is a compact, simply connected space locally isomorphic to the plane, an additional restriction on the generators appears, resulting from the properties of the space itself. This restriction suggests a completely different topological (global) character compared to the plane. It is similar in the case of the torus, which is not only a non-simply connected space but also locally isomorphic to the plane; the global character of the space is different from that of a plane, although they are both (the plane and the torus) two-dimensional spaces (the latter one, locally). We may note that the braid group for the torus has a much more complicated structure, which is associated with introducing additional generators resulting from the topological properties of this manifold.

7.6.4 The braid group for the three-dimensional Euclidean space R^3

The three-dimensional Euclidean space R^3 is a simply connected space,

$$R^3 = R \times R \times R.$$

The first homotopy group for the space R^3 is thus a trivial group,

$$\pi_1\left(R^3\right) = \pi_1\left(R \times R \times R\right) = \pi_1\left(R\right) \oplus \pi_1\left(R\right) \oplus \pi_1\left(R\right) = \varepsilon \oplus \varepsilon \oplus \varepsilon = \varepsilon.$$

Therefore, the braid group for the configuration space of N particles on the manifold R^3 has the form,

$$\pi_1\left(\left(\left(R^3\right)^N \backslash \Delta\right)/S_N\right) = \pi_1\left(Q_N\left(R^3\right)\right) = \Sigma_N\left(R^3\right),$$

and is generated by generators σ_i responsible for exchanging the i-th particles with the $i+1$-th particle. We should note that, in the space $F_N\left(R^3\right)$, every closed curve (loop) is contractible to a point by continuous transformations because in a three-dimensional space, a point cannot be encircled with a loop that would not be contracted beyond or below this point. Therefore,

$$\pi_1\left(F_N\left(R^3\right)\right) = \varepsilon.$$

According to the equation (7.53),

$$\pi_1\left(Q_N\left(R^3\right)\right)/\pi_1\left(F_N\left(R^3\right)\right) = \pi_1\left(Q_N\left(R^3\right)\right)/\varepsilon = \pi_1\left(Q_N\left(R^3\right)\right) = S_N.$$

One may observe that the equations (7.38), (7.39), and (7.47) define generators. In the case of a three-dimensional space, equation (7.47) is satisfied because a double particle exchange is topologically equivalent to describing a loop with one particle around the other and such a loop can be contracted in this space 'over' or 'under' the particle. The above properties lead to a fundamental difference between the braid group for a two-dimensional space and the braid group for a three-dimensional (or more dimensional) space. In the case of three-dimensional spaces, individual strings of the i-th and $i+1$-th particles do not tangle to form a braid. Moreover, every braid in a three-dimensional space may be disentangled. Therefore, for the case in question, only the initial and final configurations of the system are significant and not the evolution of the trajectory, meaning that,

$$\pi_1\left(Q_N\left(R^3\right)\right)/\pi_1\left(F_N\left(R^3\right)\right) = S_N.$$

For simply connected manifolds M with the dimensions $\dim M > 2$, there is an equality,

$$\pi_1\left(Q_N\left(M\right)\right) = \Sigma_N\left(M\right) = S_N.$$

This is because every loop described around a given point is contractible to the point in spaces with the dimensions $\dim M \geq 3$. In the case when $\pi_1(M) \neq \varepsilon$, if $\dim M \geq 3$, then $\Sigma_N(M) = S_N$, even if $\pi_1(Q_N(M)) \neq S_N$. The above arguments indicate that, for three or more dimensional spaces, the topological effects have no impact on $\Sigma_N(M)$, which means that the full braid group $\pi_1(Q_N(M))$ only changes when the pure braid group $\pi_1(F_N(M))$ changes as well.

7.6.5 *Braid group for a line R^1 and a circle S^1*

In the case of R^1, the space $F_N(R^1)$ is divided into separable, simply connected subspaces C_i that are numbered by the elements of the permutation group S_N, indicating that the beginning and the end of a trajectory must belong to the same subspace C_i (this statement is related to preserving the succession of particles on the line). Therefore,

$$\pi_1(Q_N(R^1)) = \pi_1(F_N(R^1)) = \varepsilon.$$

In the case of a circle S^1, which is topologically equivalent to a section with periodical boundary conditions, the notion in question becomes more complex because there is a possibility to carry out cyclical permutation of particles. The generator of the full braid group $\pi_1(Q_N(S^1))$ is the element δ, which is responsible for transferring the particle to the last location. This generator has an infinite order because subsequent powers of δ^n correspond to nonhomotopic loops; therefore,

$$\pi_1(Q_N(S^1)) = Z,$$

where Z represents an additive group of integers, whereas the group $\pi_1(F_N(S^1))$ is generated by the generator $\nu = \delta^N$. Therefore,

$$\pi_1(F_N(S^1)) = Z.$$

Although both groups $\pi_1(Q_N(S^1))$ and $\pi_1(F_N(S^1))$ correspond to the same abstract infinite group, the relation between their generators causes,

$$\pi_1(Q_N(S^1))/\pi_1(F_N(S^1)) = Z_N,$$

where Z_N is a cyclic group of order N. Similarly as in two- and three-dimensional cases, this group enumerates all possible particle arrangements on the circle resulting from the initial one.

In the case of the considered manifolds R^1 and S^1, there is no group Σ_N because it is impossible to exchange the positions of two particles without describing a non-contractible loop. The above statement indicates that the formalism of braid groups is not appropriate for one-dimensional systems [Leinaas and Myrheim (1977); Mermin (1979)].

7.7 Exact sequences for braid groups

A sequence of homomorphisms $G_1 \xrightarrow{f_1} G_2 \xrightarrow{f_2} G_3 \xrightarrow{f_3} \ldots \xrightarrow{f_{n-1}} G_n$, where G_i are groups, is exact, if

$$f_i(G_i) = Ker f_{i+1}.$$

We may observe that $F_N(M)$ (the configuration space of distinguishable identical particles with no diagonal points) causes fibration of the space $Q_N(M)$ (the configuration space of indistinguishable identical particles with no diagonal points) defined by the epimorphism h_{S_N} related to S_N operating on $F_N(M)$ [Spanier (1966); Jänich and Girlich (1984)]. Every fiber $F_N(M)_x, x \in Q_N(M)$ is a finite space (with $N!$ elements), and the fibration is locally trivial [15]. Therefore, $F_N(M)$ is a space covering $Q_N(M)$. For such fibration, the exact series of homotopy groups has the form [Hansen (1989); Spanier (1966)]:

$$\ldots \to \pi_n\left([F_N(M)]_x, y\right) \to \pi_n\left(F_N(M), y\right) \to \pi_n\left(Q_N(M), x\right) \to \ldots$$
$$\ldots \to \pi_1\left([F_N(M)]_x, y\right) \to \pi_1\left(F_N(M), y\right) \to \pi_1\left(Q_N(M), x\right) \to \quad,$$
$$\to \pi_0\left([F_N(M)]_x, y\right) \to \pi_0\left(F_N(M), y\right),$$

where M means the arcwise-connected manifold with dimension dim $M \geq 2$, $\pi_n(\Omega, \omega)$ is the n-th homotopy group for the space Ω with a base point $\omega \in \Omega$, and the points $x \in Q_N(M)$ and $y \in F_N(M)$ are arbitrarily chosen.

Because every fiber $[F_N(M)]_x$ is finite (contains $N!$ elements),

$$\pi_1\left([F_N(M)]_x, y\right) = \varepsilon$$

and

$$\pi_0\left([F_N(M)]_x, y\right) = S_N.$$

For the arcwise- connected space $F_N(M)$, one has,

$$\pi_0\left(F_N(M), y\right) = \varepsilon,$$

i.e., for every n-th homotopy group, one may omit the base point y. Thus, the sequence of homotopy has the form [Birman (1974); Hansen (1989); Spanier (1966)]:

$$\varepsilon \to \pi_1\left(F_N(M)\right) \xrightarrow{\alpha} \pi_1\left(Q_N(M)\right) \xrightarrow{\beta} S_N \to \varepsilon, \tag{7.52}$$

---

[15] h_{S_N}, which imposes locally trivial fibration if every point $x \in Q_N(M)$ has such surrounding U that make the fibration within it trivial, $F_N(M)|U = U \times F_N(M)_x$, where $F_N(M)|U$ denotes the intersection of $F_N(M)$ and the U inverse image in the projection h_{S_N}.

where ε represents a trivial single-element group, and α, β are epimorphisms.

The exactness of the sequence indicates that all generators belonging to the set L (a set of generators of the pure braid group) must also belong to the kernel of epimorphism β. Whereas, the generators $\sigma_i \in P$ (a set of generators of the exchange of neighboring particles; if M is a simply connected space, then it includes all generators of the full braid group, and the pure group is its subgroup) are transformed into the appropriate elements of the permutation group S_N. The equation (7.52) indicates that the pure braid group $\pi_1 (F_N (M))$ is a normal divisor of the full braid group $\pi_1 (Q_N (M))$ and

$$\pi_1 (Q_N (M)) / \pi_1 (F_N (M)) = S_N. \tag{7.53}$$

This means that, in the case when the manner of particle exchange is of no importance, the full braid group is a permutation group S_N.

7.8   The use of pure braid groups in classic information processing

The rich structure of braid groups in 2D creates the opportunity to use these mathematical objects to code classical information [Jacak *et al.* (2007)]. Pure braid groups, which are subgroups of the full group, seem particularly useful in this context; these include infinite groups in the case of 2D as well. Pure groups are generated by the generators l_{ij}, which satisfy relations (3.6) (paragraph 3.1.2) and refer to braids imaging particle exchanges, while preserving particle ordering (in this sense, they correspond to exchanges of identical, but distinguishable particles). If we map 0 and 1 of the binary code into two generators of the pure group of three particles on R^2 and identify the third generator with the stop character, then we can code information in a braid structure. Apart from the group properties of this new information representation (e.g., the inverse element can be useful in identifying information), the relationships between generators allow for the equivalent processing of coded information that is unavailable in binary code. This conclusion may have greater significance if we also coded classical information with more complex alphabets than those the binary code provide using pure groups for a larger number of particles (with more complex relationships for generators).

The pure braid group (in contrast to the full braid group) provides the right generator subsets for coding information, which keep their ordering; in the case of the full braid group, the generators σ_i and σ_j are commutative for $|i - j| \geq 2$ and cannot be used as basic code elements. While calculating the available number of code elements, we need to take into account the set of generators, which preserves their ordering; we may roughly assume (after considering the relationships for the generators) the linear dependence on the number of lines N. Shortening the length of the information code in a N-element alphabet, compared to the length of information coded in binary, will therefore be expressed as $\log_2(N)$. On the other hand, an increase in the number of alphabet characters (the number of lines N) makes the coding structure more complex, which increases the consumption of resources used for building subsequent connections. We should note here that presenting pure braids in 2D using entangled lines joining points of the same ordering may be identified with the structure of the connection grid in 3D, provided the initial and final points are stable. This situation gives us an opportunity to code information (in elements of the pure braid group) in such a connection network using real braid entanglements of physically realized lines. The function of resource expenditure (energy, materials) necessary to organize a network can be modeled in the power form: $g_1 = N, g_2 = N^{\frac{1}{2}}, g_3 = N^{\frac{1}{3}}, g_4 = N^{\frac{1}{4}}$. A function graph $I_i(N) = g_i(N)/(\log_2 N)$ depicted in Fig. 7.14 shows the minimal value at which coding in N alphabets is optimal in terms of the expenditure-information capacity ratio.

It is worth highlighting that braid structures represent entanglements and do not addressing the connections of these lines in the network. The advantage of the former is based on infinite information resources of appropriate braid structures in 2D compared to the finite supply provided by addressing the connections only ($N!$ for N lines). The result seems interesting [Jacak *et al.* (2007)] because it shows that the information capacity of entanglements in such a network reach a maximum at approximately 20 lines while optimizing the expenditure for the creation of the whole network for the real scaling of resource expenditure $\sim N^{1/3}$, Fig. 7.14. A small network, corresponding to an alphabet with approximately 20 elements, is therefore the most information optimal and economical in terms of the expenditure for organizing corresponding 3D connection network (in which one can code entangled braids from the 2D pure braid group). Surprisingly, there is a coincidence between this number and the number of phonemes (sounds) informatively used in most languages, which may point to a braid, not addressing structure of saving and processing linguistic information in neuronal networks.

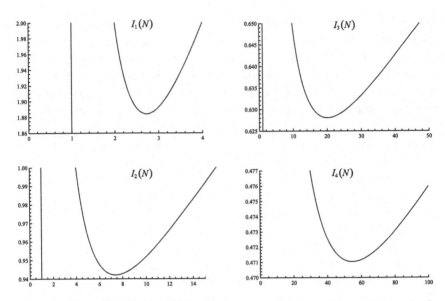

Fig. 7.14   Graphs of function $I_i(N)$, which express the ratio of expenditure to code profit for various scalings of expenditure, $g_i \sim N^{1/i}$ $i = 1, 2, 3, 4$.

Other arguments that we can cite to support the braid structure of classic information refer to its preferred two-dimensional character, corresponding to the geometry of coded information in pure braid groups for two-dimensional manifolds (only then are braid groups infinite). The two-dimensional character of information seems to be raised by the *holographic principle* [Bousso (2002); Bekenstein (1972); Hawking (1971)]. According to this principle, the entropy (hence, the information as well [Gershenfeld (2000)]) is scaled like the surface of event horizon in black holes such that the 2nd law of thermodynamics would also apply to matter collapse in a gravitational singularity [Bekenstein (1972); Hawking (1971, 1974)]. Therefore, black hole entropy was introduced [Bekenstein (1972); Hawking (1971)], $S_{BH} = \frac{A}{4}$, where A ($A = 16\pi M^2$) is the area of the event horizon of the black hole with the mass of M (units were chosen in Planck convention: $\hbar = G = c = k = 1$; then the Planck unit of area $\frac{G\hbar}{c^3} = 1$ [$= 2,59 \cdot 10^{-66}$ cm^2]). For instance, in the case of a proton, the area of the event horizon is of order of 10^{20} (in Planck units), and for Earth, 10^{41}. The finite area of the horizon limits entropy, thus limiting the number of degrees of freedom for the system, $e^{S_{BH}} = \dim H$, where H is the appropriate Hilbert space. Therefore, a black hole cannot be made of an object with too large of an entropy; it can be up to $1/4$ of the horizon area of this black hole, $A/4$, which restricts the number of degrees of freedom of a system ($\sim \ln \dim H$) and suggests that entropy (as well as information) is scaled as an area, not as a volume. Holographic models have been formulated, although not in physically justified metrics (in the hyperbolical metric *anti-de-Sitter* due to the necessity to create a border of the space for the hologram) [Suskind (1995); Maldacena and Strominger (1999); Witten (1998)], but they still underline the two-dimensional geometry of information. The overall argument for the two-dimensional character of information can be the fact that the efficient and fast information processing, which takes place on an enormously large scale in every living cell, uses information coding in deformations on the surface of properly conformed proteins (large polymer structures with sufficient surface area in atomic resolution scale; small molecules, such as H_2O, are useless in terms of information due to their small surface). Thousands of enzymatic signals, which are simultaneously being processed in cells, are likely using the most efficient information carrier, which is a deformed surface, and somehow relate to the abstract representation of the two-dimensional character of the information in braid structures (one may also map the deformed surface, after digitalizing it, into a pure braid structure in 2D).

Bibliography

Abanin, D. A., Skachko, I., Du, X., Andrei, E. Y. and Levitov, L. S. (2010). Fractional quantum Hall effect in suspended graphene: transport coefficients and electron interaction strength, *Phys. Rev. B* **81**, p. 115410.

Abrikosov, A. A. (1972). *Vvedenie v teoriu normalnych metallov* (Nauka, Moscow).

Abrikosov, A. A., Gorkov, L. P. and Dzialoshinskii, I. E. (1975). *Methods of Quantum Field Theory in Statistical Physics* (Dover Publ. Inc., Dover).

Aharonov, Y. and Bohm, D. (1959). Significance of electromagnetic potentials in quantum theory, *Phys. Rev.* **115**, p. 485.

Anderson, P. W. (1958). Absence of diffusion in certain random lattices, *Phys. Rev.* **109**, p. 1492.

Artin, E. (1947). Theory of braids, *Annals of Math.* **48**, p. 101.

Avron, J. E., Osadchy, D. and Seiler, D. (2003). A topological look at the quantum Hall effect, *Physics Today* **56**, p. 38.

Bekenstein, J. D. (1972). Black holes and the second law, *Nuovo Cim. Lett.* **4**, p. 737.

Bernevig, B. A., Hughes, T. L. and Zhang, S.-C. (2006). Quantum spin Hall effect and topological phase transition in HgTe quantum wells, *Science* **314**, p. 1757.

Berry, M. V. (1984). Quantal phase factors accompanying adiabatic changes, *Proc. R. Soc. Lond. A* **392**, p. 45.

Birkhoff, G. and Lane, S. M. (1977). *A survey of modern algebra* (Macmillan, New York).

Birman, J. S. (1969). On braid groups, *Communications on Pure and Applied Mathematics* **22**, p. 41.

Birman, J. S. (1974). *Braids, Links and Mapping Class Groups* (Princeton UP, Princeton).

Bishara, W. and Nayak, C. (2009). Effect of Landau level mixing on the effective interaction between electrons in the fractional quantum Hall regime, *Phys. Rev. B* **80**, p. 121302.

Bloom, D. M. (1979). *Linear Algebra and Geometry* (Cambridge University Press, Cambridge).

Bolotin, K. I., Ghahari, F., Shulman, M. D., Störmer, H. L. and Kim, P. (2009). Observation of the fractional quantum Hall effect in graphene, *Nature* **462**, p. 196.

Bolotin, K. I., Sikes, K. J., Jiang, Z., Klima, M., Fudenberg, G., Hone, J., Kim, P. and Störmer, H. L. (2008). Ultrahigh electron mobility in suspended graphene, *Solid State Comm.* **146**, p. 351.

Bousso, R. (2002). The holographic principle, *Rev. Mod. Phys.* **74**, p. 825.

Bouwmeester, D., Ekert, A. and Zeilinger, A. (2000). *The Physics of Quantum Information* (Springer Verlag, Berlin).

Castro Neto, A. H., Guinea, F., Peres, N. M. R., Novoselov, K. S. and Geim, A. K. (2009). The electronic properties of graphene, *Rev. Mod. Phys.* **81**, p. 109.

Chaichian, M. and Demichev, A. (2001a). *Path Integrals in Physics Volume I Stochastic Processes and Quantum Mechanics* (IOP Publishing Ltd, Bristol; Philadelphia).

Chaichian, M. and Demichev, A. (2001b). *Path Integrals in Physics Volume II Quantum Field Theory, Statistical Physics and other Modern Applications* (IOP Publishing Ltd, Bristol; Philadelphia).

Cheng, P., Song, C., Zhang, T., Zhang, Y., Wang, Y., Jia, J.-F., Wang, J., Wang, Y., Zhu, B.-F., Chen, X., Ma, X., He, K., Wang, L., Dai, X., Fang, Z., Xie, X., Qi, X.-L., Liu, C.-X., Zhang, S.-C. and Que, Q.-K. (2010). Landau quantization of topological surface states in Bi_2Se_3, *Phys. Rev. Lett.* **105**, p. 076801.

Chern, S. S. and Simons, J. (1974). Characteristic forms and geometric invariants, *Annals of Math.* **99**, p. 48.

Coxeter, H. S. and Moser, W. O. (1984). *Generators and relations for discrete groups* (Springer-Verlag, Berlin).

Dahal, H. P., Joglekar, Y. N., Bedell, K. S. and Balatsky, A. V. (2006). Absence of Wigner crystallization in graphene, *Phys. Rev. B* **74**, p. 233405.

Das Sarma, S. and Pinczuk, A. (1997). *Perspectives in Quantum Hall Effects: Novel Quantum Liquids in Low-Dimensional Semiconductor Structures* (Wiley, New York).

Dean, C. R., Young, A. F., Cadden-Zimansky, P., Wang, L., Ren, H., Watanabe, K., Taniguchi, T., Kim, P., Hone, J. and Shepard, K. L. (2011). Multicomponent fractional quantum Hall effect in graphene, *Nature Physics* **7**, p. 693.

Du, X., Skachko, I., Duerr, F., Luican, A. and Andrei, E. Y. (2009). Fractional quantum Hall effect and insulating phase of Dirac electrons in graphene, *Nature* **462**, p. 192.

Dyson F. quoted in Schrieffer J. R. (1983). *Theory of superconductivity* (Benjamin-Cummings, Reading, MA), p. 42.

Einarsson, T. (1990). Fractional statistics on a torus, *Phys. Rev. Lett.* **64**, p. 1995.

Eisenstein, J. P., Cooper, K. B., Pfeiffer, L. N. and West, K. W. (2002). Insulating and fractional quantum Hall states in the first excited landau level, *Phys. Rev. Lett.* **88**, p. 076801.

Eliutin, P. W. and Krivchenkov, W. D. (1976). *Kantovaja miechanika* (Nauka, Moscow).

English, J. H., Gossard, A. C., Störmer, H. L. and Baldwin, K. W. (1987). GaAs structures with electron mobility of 5×10^6 cm^2/Vs, *Appl. Phys. Lett.* **50**, p. 1826.

Fetter, A. L., Hanna, C. B. and Laughlin, R. B. (1989). Random-phase approximation in the fractional-statistics gas, *Phys. Rev. B* **39**, p. 9679.

Feynman, R. P. and Hibbs, A. R. (1964). *Quantum Mechanics and Path Integrals* (McGraw-Hill, New York).

Fleischmann, R. (1996). *Nichtlineare Dynamic in Halbleiter-Nanostrukturen*, Ph.D. thesis, Frankfurt University.

Geim, A. K. and MacDonald, A. H. (2007). Graphene: exploring carbon flatland, *Physics Today* **60**, p. 35.

Gemelke, N. (2007). *Quantum degenerate atomic gases in controlled optical lattice potentials*, Ph.D. thesis, Stanford University. Dept. of Physics.

Gershenfeld, N. (2000). *The Physics of Information Technology* (Cambridge U.P., Cambridge).

Girvin, S. M. and MacDonald, A. H. (1987). Off-diagonal long-range order, oblique confinement, and the fractional quantum Hall effect, *Phys. Rev. Lett.* **58**, p. 1252.

Girvin, S. M., MacDonald, A. H. and Platzman, P. M. (1985). Collective-excitation gap in the fractional quantum Hall effect, *Phys. Rev. Lett.* **54**, p. 581.

Goerbig, M. O. (2011). Electronic properties of graphene in a strong magnetic field, *Review of Modern Physics* **83**, p. 1193.

Goerbig, M. O. and Regnault, N. (2007). Analysis of a SU(4) generalization of Halperins wave function as an approach towards a SU(4) fractional quantum Hall effect in graphene sheets, *Phys. Rev. B* **75**, p. 241405.

Greiter, M., Wen, X. G. and Wilczek, F. (1992). Paired Hall states, *Nuclear Phys. B* **374**, p. 567.

Grimes, C. C. and Adams, G. (1979). Evidence for liquid-to-crystal phase transition in a classical, two-dimensional sheet of electrons, *Phys. Rev. Lett.* **42**, p. 795.

Haldane, F. D. M. (1983). Fractional quantization of the Hall effect: a hierarchy of incompressible quantum fluid states, *Phys. Rev. Lett.* **51**, p. 605.

Haldane, F. D. M. (1988). Model of quantum Hall effect without Landau levels: condensed matter realization of the 'parity anomaly', *Phys. Rev. Lett.* **61**, p. 2015.

Halperin, B. I. (1983). Theory of the quantized Hall conductance, *Helv. Phys. Acta* **56**, p. 75.

Halperin, B. I., Lee, P. A. and Read, N. (1993). Theory of the half-filled Landau level, *Phys. Rev. B* **47**, p. 7312.

Hansen, V. L. (1989). *Braids and coverings* (Cambridge UP, Cambridge).

Hasan, M. Z. and Kane, C. L. (2010). Colloquium: Topological insulators, *Rev. Mod. Phys.* **82**, 3045.

Hasan, M. Z. and Moore, J. E. (2011). Three-dimensional topological insulators, *Annu. Rev. Condens. Matter Phys.* **2**, p. 55.

Hatcher, A. (2002). *Algebraic topology* (Cambridge University Press, Cambridge).

Hawking, S. W. (1971). Gravitational radiation from colliding black holes, *Phys. Rev. Lett.* **26**, p. 1344.

Hawking, S. W. (1974). Black hole explosions? *Nature* **248**, p. 30.

Heinonen, O. (1998). *Composite Fermions* (World Scientific, Singapore).

Imbo, T. D., Imbo, C. S. and Sudarshan, C. S. (1990). Identical particles, exotic statistics and braid groups, *Phys. Lett. B* **234**, p. 103.

Imbo, T. D. and Sudarshan, E. C. G. (1988). Inequivalent quantizations and fundamentally perfect spaces, *Phys. Rev. Lett.* **60**, p. 481.

Jacak, J. and Jacak, L. (2010). Recovery of Laughlin correlations with cyclotron braids, *Euorophysics Lett.* **92**, p. 60002.

Jacak, J., Jóźwiak, I. and Jacak, L. (2007). Application of braid groups for information processing, in H. R. Arabnia and P. L. Zou (eds.), *Proc. of the 2007 International Conference on Foundation of Computer Science* (CSREA Press), p. 344.

Jacak, J., Jóźwiak, I. and Jacak, L. (2009a). New implementation of composite fermions in terms of subgroups of a braid group, *Phys. Lett. A* **374**, p. 346.

Jacak, J., Jóźwiak, I. and Jacak, L. (2010a). Composite fermions in braid group terms, *Open Sys. and Inf. Dyn.* **17**, p. 1.

Jacak, J., Jóźwiak, I., Jacak, L. and Wieczorek, K. (2010b). Cyclotron braid group structure for composite fermions, *J. Phys: Cond. Matt.* **22**, p. 355602.

Jacak, J., Jóźwiak, I., Jacak, L. and Wieczorek, K. (2011). Cyclotron braid group approach to Laughlin correlations, *Adv. Theor. Math. Phys.* **15**, p. 1.

Jacak, L. (1988). *Nonlinear topics in Fermi liquid theory* (Oficyna Wyd. PWr, Wroclaw).

Jacak, L., Krasnyj, J., Jacak, W., Gonczarek, R. and Machnikowski, P. (2005). Unavoidable decoherence in semiconductor quantum dots, *Phys. Rev. B* **72**, p. 245309.

Jacak, L., Sitko, P., Wieczorek, K. and Wójs, A. (2003). *Quantum Hall Systems: Braid groups, composite fermions, and fractional charge* (Oxford UP, Oxford).

Jacak, W., Krasnyj, J., Jacak, L. and Gonczarek, R. (2009b). *Dekoherencja orbitalnych i spinowych stopni swobody w kropkach kwantowych* (Oficyna Wydawnicza PWR, Wroclaw).

Jain, J. K. (1989). Composite-fermion approach for the fractional quantum Hall effect, *Phys. Rev. Lett.* **63**, p. 199.

Jain, J. K. (2007). *Composite Fermions* (Cambridge UP, Cambridge).

Jänich, K. and Girlich, H. J. (1984). *Topology* (Springer-Verlag, New York, Berlin).

Josephson, B. D. (1974). The discovery of tunnelling supercurrents, *Rev. Mod. Phys.* **46**, p. 251.

Kane, C. L. and Mele, E. J. (2005). Quantum spin Hall effect in graphene, *Phys. Rev. Lett.* **95**, p. 226801.

Kato, Y. K., Myers, R. C., Gossard, A. C. and Awschalom, D. D. (2004). Observation of the spin Hall effect in semiconductors, *Science* **306**, 5703, p. 1910.

Kim, J., Kim, J. and Jhi, S.-H. (2010). Prediction of topological insulating behavior in crystalline Ge-Sb-Te, *Phys. Rev. B* **82**, p. 201312(R).

Kitaev, A. Y. (2003). Fault-tolerant quantum computation by anyons, *Annals of Phys.* **303**, p. 2.

Kivelson, S., Kallin, C., Arovas, D. P. and Schrieffer, J. R. (1986). Cooperative ring exchange theory of the fractional quantized Hall effect, *Phys. Rev. Lett.* **56**, p. 873.

König, M., Wiedmann, S., Brüne, C., Roth, A., Buhmann, H., Molenkamp, L. W., Qi, X. and Zhang, S. (2007). Quantum spin Hall insulator state in HgTe quantum wells, *Science* **318**, p. 766.

Laidlaw, M. G. and DeWitt, C. M. (1971). Feynman functional integrals for systems of indistinguishable particles, *Phys. Rev. D* **3**, p. 1375.

Landau, L. D. and Lifshitz, E. M. (1972). *Quantum Mechanics: Non-relativistic Theory* (Nauka, Moscow).

Lang, S. (2002). *Algebra* (Springer-Verlag, New York).

Laughlin, R. B. (1983a). Anomalous quantum Hall effect: an incompressible quantum fluid with fractionally charged excitations, *Phys. Rev. Lett.* **50**, p. 1395.

Laughlin, R. B. (1983b). Quantized motion of three two-dimensional electrons in a strong magnetic field, *Phys. Rev. B* **27**, p. 3383.

Lee, D. H. (1998). Neutral fermions at filling factor v = 1/2, *Phys. Rev. Lett.* **80**, p. 4745.

Lee, D. H., Baskaran, G. and Kivelson, S. (1987). Generalized cooperative-ring-exchange theory of the fractional quantum Hall effect, *Phys. Rev. Lett.* **59**, p. 2467.

Leinaas, J. M. and Myrheim, J. (1977). On the theory of identical particles, *Nuovo Cimmento* **37B**, p. 1.

Levesque, D., Weiss, J. J. and MacDonald, A. H. (1984). Crystallization of the incompressible quantum-fluid state of a two-dimensional electron gas in a strong magnetic field, *Phys. Rev. B* **30**, p. 1056.

Levin, M. and Stern, A. (2009). Fractional Topological Insulators, *Phys. Rev. Lett.* **103**, p. 196803.

Lopez, A. and Fradkin, E. (1991). Fractional quantum Hall effect and Chern-Simons gauge theories, *Phys. Rev. B* **44**, p. 5246.

Luhman, D. R., Pan, W., Tsui, D. C., Pfeiffer, L. N., Baldwin, K. W. and West, K. W. (2008). Observation of a fractional quantum Hall state at v=1/4 in a wide GaAs quantum well, *Phys. Rev. Lett.* **101**, p. 266804.

MacClure, J. W. (1956). Diamagnetism of graphite, *Phys. Rev.* **104**, p. 666.

Maki, K. and Zotos, X. (1983). Static and dynamic properties of a two-dimensional Wigner crystal in a strong magnetic field, *Phys. Rev. B* **28**, p. 4349.

Maldacena, J. and Strominger, A. (1999). Semiclassical decay of near extremal fivebranes, *J. High Energy Phys.* **12**, p. 008.

Mandal, S. S., Peterson, M. R. and Jain, J. K. (2003). Two-dimensional electron system in high magnetic fields: Wigner crystal versus composite-fermion liquid, *Phys. Rev. Lett.* **90**, p. 106403.

Mani, R. G., Smet, J. H., v.Klitzing, K., Narayanamurti, V., Johnson, W. B. and Umansky, V. (2002). Zero-resistance states induced by electromagnetic-wave excitation in GaAs/AlGaAs heterostructures, *Nature* **420**, p. 646.

Mermin, N. (1979). The topological theory of defects in ordered media, *Rev. Mod. Phys.* **51**, p. 591.

Möller, G. and Simon, S. H. (2008). Paired composite-fermion wave functions, *Phys. Rev. B* **77**, p. 075319.

Moore, G. and Read, N. (1991). Nonabelions in the fractional quantum Hall effect, *Nuclear Phys. B* **360**, p. 362.

Moore, J. E. (2009). Nature physics, *Topological insulators: The next generation* **5**, p. 378.

Morinari, T. (2000). Composite fermion pairing theory in single-layer systems, *Phys. Rev. B* **62**, p. 15903.

Nakahara, M. (1990). *Geometry, Topology and Physics* (Adam Hilger, Bristol).

Nayak, C., Simon, S. H., Stern, A., Freedman, M. and Das Sarma, S. (2008). Nonabelian anyons and topological quantum computation, *Rev. Mod. Phys.* **80**, p. 1083.

Nielsen, M. A. and Chuang, I. L. (2000). *Quantum Computation and Quantum Information* (Cambridge UP, Cambridge).

Noh, H.-J., Jeong, J., Cho, E.-J., Lee, H.-K. and Kim, H.-D. (2011). Persistence of surface states despite impurities in the surface of topological insulators, *Europhysics Lett.* **96**, p. 47002.

Novoselov, K. S., Geim, A. K., Morozov, S. V., Jiang, D., Katsnelson, M. I., Grigorieva, I. V., Dubonos, S. V. and Firsov, A. A. (2005). Two-dimensional gas of massles Dirac fermions in graphene, *Nature* **438**, p. 197.

Olchovskii, I. I. (1974). *Kurs teoreticeskoj mechaniki dlia fizikov* (Izdat. Mosk. Univ., Moscow).

Palmer, R. N. and Jaksch, D. (2006). High-field fractional Hall effect in optical lattices, *Phys. Rev. Lett.* **96**, p. 180407.

Palmer, R. N., Klein, A. and Jaksch, D. (2008). Optical lattice quantum Hall effect, *Phys. Rev. A* **78**, p. 013609.

Pan, W., Störmer, H. L., Tsui, D. C., Pfeiffer, L. N., Baldwin, K. W. and West, K. W. (2003). Fractional quantum hall effect of composite fermions, *Phys. Rev. Lett.* **90**, p. 016801.

Papadopoulos, G. J. and Devreese, J. T. (1978). *Path Integrals and Their Applications in Quantum, Statistical, and Solid State Physics* (Plenum Press, New York).

Papić, Z., Goerbig, M. O. and Regnault, N. (2009a). Theoretical expectations for a fractional quantum Hall effect in graphene, *Solid State Comm.* **149**, p. 1056.

Papić, Z., Goerbig, M. O. and Regnault, N. (2010). Atypical fractional quantum Hall effect in graphene at fillig factor 1/3, *Phys. Rev. Lett.* **105**, p. 176802.

Papić, Z., Möller, G., Milovanović, M. V., Regnault, N. and Goerbig, M. O. (2009b). Fractional quantum Hall state at v = 1/4 in a wide quantum well, *Phys. Rev. B* **79**, p. 245325.

Pasquier, V. and Haldane, F. D. M. (1998). A dipole intepretation of v = 1/2 state, *Nuclear Phys. B* **516**, p. 719.

Pfeiffer, L. and West, K. W. (2003). The role of MBE in recent quantum Hall effect physics discoveries, *Physica E* **20**, p. 57.

Polyakov, A. M. (1974). Particle spectrum in quantum field theory, *Zhurn. Eksp. Teor. Fiz., Pis. Red.* **20**, p. 430.

Prange, R. E. and Girvin, S. M. (1990). *The Quantum Hall Effect* (Springer Verlag, New York).

Preskill, J. (2004). *Topological Quantum Computation* (California Inst. Tech.), Lecture Notes for Physics vol. 219.

Prodan, E. (2011). Disordered topological insulators: a non-commutative geometry perspective, *J. Phys. A: Math. Theor.* **44**, p. 113001.

Qi, X.-L. (2011). Generic wave-function description of fractional quantum anomalous Hall states and fractional topological insulators, *Phys. Rev. Lett.* **107**, p. 126803.

Qi, X.-L. and Zhang, S.-C. (2010). The quantum spin Hall effect and topological insulators, *Physics Today* **63**, p. 33.

Qi, X.-L. and Zhang, S.-C. (2011). Topological insulators and superconductors, *Rev. Mod. Phys.* **83**, p. 1057.

Rajaraman, R. and Sondhi, S. L. (1996). A field theory for the Read operator, *Int. J. Mod. Phys. B* **10**, p. 793.

Read, N. (1989). Order parameter and Ginzburg-Landau theory for the fractional quantum Hall effect, *Phys. Rev. Lett.* **62**, p. 86.

Read, N. (1994). Theory of the half-filled Landau level, *Semicond. Sci. Technol.* **9**, p. 1859.

Read, N. (1998). Lowest-Landau-level theory of the quantum Hall effect: The Fermi-liquid-like state of bosons at filling factor one, *Phys. Rev. B* **58**, p. 16262.

Read, N. and Green, D. (2000). Paired states of fermions in two dimensions with breaking of parity and time-reversal symmetries and the fractional quantum Hall effect, *Phys. Rev. B* **61**, p. 10267.

Rezayi, E. H. and Read, N. (1994). Fermi-liquid-like state in a half-filled Landau level, *Phys. Rev. Lett.* **72**, p. 900.

Rezayi, E. H. and Read, N. (2009). Non-abelian quantized Hall states of electrons at filling factors 12/5 and 13/5 in the first excited Landau level, *Phys. Rev. B* **79**, p. 075306.

Rezayi, E. H. and Simon, S. H. (2011). Breaking of particle-hole symmetry by Landau level mixing in the $\nu=5/2$ quantized Hall state, *Phys. Rev. Lett.* **106**, p. 116801.

Roy, R. (2009). Topological phases and the quantum spin Hall effect in three dimensions, *Phys. Rev. B* **79**, p. 195322.

Ryder, L. H. (1996). *Quantum Field Theory, 2nd ed.* (Cambridge University Press, Cambridge).

Sato, T., Segawa, K., Guo, H., Sugawara, K., Souma, S., Takahashi, T. and Ando, Y. (2010). Direct evidence for the Dirac-cone topological surface states in the ternary chalcogenide TlBiSe2, *Phys. Rev. Lett.* **105**, p. 136802.

Sbeouelji, T. and Meskini, N. (2001). Stability of the fractional quantum Hall effect in higher Landau levels for composite fermions carrying four vortices, *Phys. Rev. B* **64**, p. 193305.

Serre, J.-P. (1977). *Linear Representations of Finite Groups* (Springer-Verlag, New York).

Shabani, J., Gokmen, T., Chiu, Y. T. and Shayegan, M. (2009a). Evidence for developing fractional quantum Hall states at even denominator 1/2 and 1/4 fillings in asymmetric wide quantum wells, *Phys. Rev. Lett.* **103**, p. 256802.

Shabani, J., Gokmen, T. and Shayegan, M. (2009b). Correlated states of electrons in wide quantum wells at low fillings: the role of charge distribution symmetry, *Phys. Rev. Lett.* **103**, p. 046805.

Shankar, R. and Murthy, G. (1997). Towards a field theory of fractional quantum Hall states, *Phys. Rev. Lett.* **79**, p. 4437.

Sheng, D. N., Gu, Z.-C., Sun, K. and Sheng, L. (2011). Fractional quantum Hall effect in the absence of Landau levels, *Nature Comm.* **2**, p. 389.

Simon, S. H., Rezayi, E. H. and Cooper, N. R. (2007). Pseudopotentials for multiparticle interactions in the quantum Hall regime, *Phys. Rev. B* **75**, p. 195306.

Skachko, I., Du, X., Duerr, F., Luican, A., Abanin, D. A., Levitov, L. S. and Andrei, E. Y. (2010). Fractional quantum Hall effect in suspended graphene probed with two-terminal measurements, *Phil. Trans. R. Soc. A* **368**, p. 5403.

Sørensen, A. S., Demler, E. and Lukin, M. D. (2005). Fractional quantum Hall states of atoms in optical lattices, *Phys. Rev. Lett.* **94**, p. 086803.

Spanier, E. (1966). *Algebraic topology* (Springer-Verlag, Berlin).

Stanescu, T. D., Galitski, V. and Das Sarma, S. (2010). Topological states in two-dimensional optical lattices, *Phys. Rev. A* **82**, p. 013608.

Sudarshan, E. C. G., Imbo, T. D. and Govindarajan, T. R. (1988). Configuration space topology and quantum internal symmetries, *Phys. Lett. B* **213**, p. 471.

Sun, K., Gu, Z., Katsura, H. and Das Sarma, S. (2011). Nearly flatbands with nontrivial topology, *Phys. Rev. Lett.* **106**, p. 236803.

Sun, K., Yao, H., Fradkin, E. and Kivelson, S. A. (2009). Topological insulators and nematic phases from sponteneous symmetry breaking in 2D Fermi systems with a quadratic band crossing, *Phys. Rev. Lett.* **103**, p. 046811.

Suskind, L. (1995). The world as a hologram, *J. Math. Phys.* **36**, p. 6377.

t'Hooft, G. (1974). Magnetic monopoles in unified gauge theories, *Nuclear Phys. B* **79**, p. 276.

t'Hooft, G. (1976). Computation of the quantum effects due to a four-dimensional pseudoparticle, *Phys. Rev. D* **14**, p. 3432.

Thouless, D. J., Kohmoto, M., Nightingale, M. P. and den Nijs, M. (1982). Quantized Hall conductance in a two-dimensional periodic potential, *Phys. Rev. Lett.* **49**, p. 405.

Tsui, D. C., Störmer, H. L. and Gossard, A. C. (1982). Two-dimensional magnetotransport in the extreme quantum limit, *Phys. Rev. Lett.* **48**, p. 1559.

von Klitzing, K., Dorda, G. and Pepper, M. (1980). New method for high-accuracy determination of the fine-structure constant based on quantized Hall resistance, *Phys. Rev. Lett.* **45**, p. 494.

Wallace, P. R. (1947). The band theory of graphite, *Phys. Rev.* **71**, p. 622.

Wang, Z., Qi, X.-L. and Zhang, S.-C. (2010). Topological order parameters for interacting topological insulators, *Phys. Rev. Lett.* **105**, p. 256803.

Wigner, E. P. (1934). On the interaction of electrons in metals, *Phys. Rev.* **46**, p. 1002.

Wilczek, F. (1990). *Fractional Statistics and Anyon Superconductivity* (World Scientific, Singapore).

Willett, R. L., Eisenstein, J. P., Störmer, H. L., Tsui, D. C., Gossard, A. C. and English, J. H. (1987). Observation of an even-denominator quantum number in the fractional quantum Hall effect, *Phys. Rev. Lett.* **59**, p. 1776.

Willett, R. L., Ruel, R. R., West, K. W. and Pfeiffer, L. N. (1993). Experimental demonstration of a Fermi surface at one-half filling of the zeroth Landau level, *Phys. Rev. Lett.* **71**, p. 3846.

Willett, R. L., Störmer, H. L., Tsui, D. C., Pfeiffer, L. N., West, K. W. and Baldwin, K. W. (1988). Termination of the series of fractional quantum Hall states at small filling factors, *Phys. Rev. B* **38**, p. 7881.

Willett, R. L., West, K. W. and Pfeiffer, L. N. (1995). Apparent inconsistency of observed composite fermion geometric resonances and measured effective mass, *Phys. Rev. Lett.* **75**, p. 2988.

Witten, E. (1995). Chern-Simons gauge theory as a string theory, *Prog. Math.* **133**, p. 637.

Witten, E. (1998). Gauge theory correlators from non-critical string theory, *Adv. Theor. Math. Phys.* **2**, p. 253.

Wu, C., Bergman, D., Balents, L. and Das Sarma, S. (2007). Flat bands and Wigner crystalization in honeycomb optical lattice, *Phys. Rev. Lett.* **99**, p. 070401.

Wu, Y. S. (1984). General theory for quantum statistics in two dimensions, *Phys. Rev. Lett.* **52**, p. 2103.

Xia, J. S., Pan, W., Vicente, C. L., Adams, E. D., Sullivan, N. S., Störmer, H. L., Tsui, D. C., Pfeiffer, L. N., Baldwin, K. W. and West, K. W. (2004). Electron correlation in the second Landau level: a competition between many nearly degenerate quantum phases, *Phys. Rev. Lett.* **93**, p. 176809.

Xia, Y., Qian, D., Hsieh, D., Wray, L., Pal, A., Lin, H., Bansil, A., Grauer, D., Hor, Y. S., Cava, R. J. and Hasan, M. Z. (2009). Observation of a large-gap topological-insulator class with a single Dirac cone on the surface, *Nature Phys.* **5**, 6, p. 398.

Yakovenko, V. M. (1990). Chern-Simons term and **n** field in Haldane's model for quantum Hall effect without Landau levels, *Phys. Rev. Lett.* **65**, p. 251.

Yang, K. (2007). Spontaneous symmetry breaking and quantum Hall effect in graphene, *Sol. State Comm.* **143**, p. 27.

Yang, K., Haldane, F. D. M. and Rezayi, E. H. (2001). Wigner crystal in a lowest Landau level at low filling factors, *Phys. Rev. B* **64**, p. 0831301(R).

Yannouleas, C., Romanovsky, I. and Landman, U. (2010). Edge and bulk components of lowest-Landau-level orbitals, correlated fractional quantum Hall effect incompressible states, and insulating behavior of finite graphene samples, *Phys. Rev. B* **82**, p. 125419.

Zhang, M., Huang, H., Zhang, C. and Wu, C. (2011). Quantum anomalous Hall states in the p-orbital honeycomb optical lattice, *Phys. Rev. A* **83**, p. 023615.

Zhang, S.-C., Hansson, T. H. and Kivelson, S. (1989). Effective-field-theory model for the fractional quantum Hall effect, *Phys. Rev. Lett.* **62**, p. 82.

Zhang, Y., Jiang, Z., Small, J. P., Purewal, M. S., Tan, Y.-W., Fazlollahi, M., Chudov, J. D., Jaszczak, J. A., Störmer, H. L. and Kim, P. (2006). Landau-level splitting in graphene in high magnetic fields, *Phys. Rev. Lett.* **96**, p. 136806.

Zhang, Y., Tan, Y.-W., Störmer, H. L. and Kim, F. (2005). Experimental observation of the quantum Hall effect and Berry's phase in graphene, *Nature* **438**, p. 201.

Zudov, M. A., Du, R. R., Pfeiffer, L. N. and West, K. W. (2003). Evidence for a new dissipationless effect in 2D electronic transport, *Phys. Rev. Lett.* **90**, p. 046807.

Zudov, M. A., Du, R. R., Simmons, J. A. and Reno, J. (2001). Shubnikov-de Haas-like oscillations in millimeterwave photoconductivity in a high-mobility two-dimensional electron gas, *Phys. Rev. B* **64**, p. 201331.

Index